Otolith Atlas from the Persian Gulf and the Oman Sea Fishes

Zahra Sadighzadeh
Víctor M. Tuset

Otolith Atlas from the Persian Gulf and the Oman Sea Fishes

LAP LAMBERT Academic Publishing

Impressum / Imprint
Bibliografische Information der Deutschen Nationalbibliothek: Die Deutsche Nationalbibliothek verzeichnet diese Publikation in der Deutschen Nationalbibliografie; detaillierte bibliografische Daten sind im Internet über http://dnb.d-nb.de abrufbar.
Alle in diesem Buch genannten Marken und Produktnamen unterliegen warenzeichen-, marken- oder patentrechtlichem Schutz bzw. sind Warenzeichen oder eingetragene Warenzeichen der jeweiligen Inhaber. Die Wiedergabe von Marken, Produktnamen, Gebrauchsnamen, Handelsnamen, Warenbezeichnungen u.s.w. in diesem Werk berechtigt auch ohne besondere Kennzeichnung nicht zu der Annahme, dass solche Namen im Sinne der Warenzeichen- und Markenschutzgesetzgebung als frei zu betrachten wären und daher von jedermann benutzt werden dürften.

Bibliographic information published by the Deutsche Nationalbibliothek: The Deutsche Nationalbibliothek lists this publication in the Deutsche Nationalbibliografie; detailed bibliographic data are available in the Internet at http://dnb.d-nb.de.
Any brand names and product names mentioned in this book are subject to trademark, brand or patent protection and are trademarks or registered trademarks of their respective holders. The use of brand names, product names, common names, trade names, product descriptions etc. even without a particular marking in this works is in no way to be construed to mean that such names may be regarded as unrestricted in respect of trademark and brand protection legislation and could thus be used by anyone.

Coverbild / Cover image: www.ingimage.com

Verlag / Publisher:
LAP LAMBERT Academic Publishing
ist ein Imprint der / is a trademark of
AV Akademikerverlag GmbH & Co. KG
Heinrich-Böcking-Str. 6-8, 66121 Saarbrücken, Deutschland / Germany
Email: info@lap-publishing.com

Herstellung: siehe letzte Seite /
Printed at: see last page
ISBN: 978-3-8454-0687-9

Copyright © 2012 AV Akademikerverlag GmbH & Co. KG
Alle Rechte vorbehalten. / All rights reserved. Saarbrücken 2012

Table of Contents

Introduction 1

Material and Methods 4

Morphological descriptions 11

References 55

Acknowledgements 58

OTOLITH ATLAS

FROM THE PERSIAN GULF AND OMAN SEA FISHES

Zahra Sadighzadeh[1], Víctor M. Tuset[2], Mohammad R. Dadpour[3], José L. Otero-Ferrer[2] & Antoni Lombarte[2]

[1]Marine Biology Department, Faculty of Marine Science & Technology, Science and Research Branch, Islamic Azad University, Tehran, Iran;
[2]Instituto de Ciencies del Mar (CSIC), Barcelona, Catalonia, Spain;
[3]Department of Horticultural Sciences, Faculty of Agriculture, University of Tabriz, Tabriz, Iran

Introduction

The Persian Gulf is a relatively young sea that originated about 16,000 BP, attaining the current level 6,000 years BP, during the Holocene (Sheppard et al. 1992). By 14,000 BP the Hormuz Strait has opened up as a narrow waterway and the flooding of the lowlands to the west begins, first with the flooding of the Eastern Basin by marine water soon after 13,000 BP, and establishing the present fish fauna (Beech 2004). Thus, the Persian Gulf is became in a semi-enclosed marine system connected to the Indian Ocean through the Arabian Sea. The Oman Sea is generally included as a branch of the Persian Gulf, not as an arm of the Arabian Sea. The surface water flows into the Gulf in the northern part of the Strait of Hormuz as a wedge of less saline water that penetrates deep into the Gulf along the Iranian coast (Brewer and Dyrssen 1985; Reynolds 1993). Evaporation level is far greater than the combined rainfall and river discharge, leading both to an inverse estuarine circulation and a counterclockwise circulation, becoming to this ecosystem in extreme marine environment (Chao et al. 1992; Kämpf and Sadrinasab 2006).

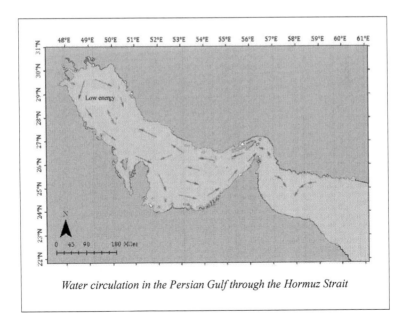

Water circulation in the Persian Gulf through the Hormuz Strait

Considering the biodiversity of the Gulf and Oman Sea, it can be stated that the wildlife in the region is unique, corals, dugongs, birds, sea turtles, oysters, jellyfish, sea star, sea cucumber, anemones, crabs and many other species even whales and whale sharks live in this marine ecosystem. A report by the United Nations University Institute for Water, Environment and Health shows that "the unprecedented scale of economic growth in the region and the significant development along the coastlines of many Gulf countries is putting pressure on coastal ecosystems. This has contributed to considerable degradation of natural habitats, including mangroves, seagrass beds, coral reefs and marine life. Nowadays, many valuable marine fauna and flora inhabit are near extinction or at serious environmental risk as mangroves, which act as nurseries for many marine species, and corals, which support multitude of marine species and whose health directly reflects the health of the Gulf. The consequences could be disaster for the fishing activity of region hence over 80% depend on, directly or indirectly, the coral reef for their survival. Reefs have been rapidly declining worldwide over the past 50 years, however the Gulf, with only about 1.5% of the worlds reefs, is one of the most grievously affected regions; over 70% of its original 3,800 km2 reef cover may be considered lost and a further 27% threatened or at critical stages of degradation (Van Lavieren 2011).

Next to the oil, fisheries represent the second most important natural resource. Reported catches in southern waters of Iran shows a generally increasing trend over the period 1993–2008. Total catch in southern waters is 342,000 tonnes, and a major part of the catch increase is attributed to greater landings of tuna, sardines and hairtail. The value of the southern catch in 2008 was estimated at USD$ 423 million with an export value of USD$60 million (Valinassab 2006, 2010). The Gulf is one of the most productive bodies of water in the world (Sheppard 1993), where deeper waters of the northern and eastern Gulf have higher species richness than waters in the southern region (Price et al. 1993). According to estimates made by the U.N. Food and Agriculture Organization (FAO), potential fishery resources in the Gulf amount to 550,000 tonnes annually, some eight times greater than in the Gulf of Oman (Kardovani 1995).

Bony fishes are the major inhabitants the region and they are preys and depredators of other fishes or marine organism. The management and assessment of the natural resources require of the knowledge of trophic networks. The identification and classification of species in the digestive content of marine organism is a highlight research line in system ecology. In the case of boy fishes their identification is performed using, mainly, one bone localized in inner ear of fishes called "*sagittal* otolith", which together another two otoliths, *lapilli* and *asterisci*, carry out the vestibular (balance) and acoustic (sound detection) function (Platt and Popper 1981; Popper and Fay 1993). They are acellular concretions of calcium carbonate another inorganic salts, which develop over a protein matrix (Carlström 1963; Blacker 1969; Degens et al. 1969; Cermeño et al. 2006) which shapes have taxonomic value (Fitch and Brownell, 1968; Schmidt, 1969; Nolf, 1985). For that reason, otolith catalogues has been published around word: Schmidt (1968), east Atlantic; Morrow (1977, 1979), North American Pacific coast; Nolf (1985), of global scope, including fossil otoliths; Härkönen (1986), North Sea fishes; Hecht (1987) and Smale et al. (1995), South African fish species;

Williams and McEldowney (1990), Australian Antarctic region; Rivaton and Bourret (1999), Indo Pacific region; García-Godos and Naveda (2001) Peruvian waters; Volpedo and Echeverría (2000), Argentinian fishes; Assis (2000, 2004), Portuguese coastal, estuarine and freshwater species; Campana (2004), western Atlantic species; Furlani et al. (2007) Australian temperate; and finally Tuset et al. (2008). The unification of criteria for the otolith description was made by Tuset et al. (2008). Therefore, it book does not go into detail about this aspect and we recommend the reader of Tuset et al. (2008) for more details. Many of these otoliths can be noted in AFORO website (http://www.cmima.csic.es/aforo/), a data-base associated with a Automatic Species Identification (Parisi-Barad et al. 2010).

In 1973 was created the ROPME (Regional Organization for the Protection of the Marine Environment) founded by eight coastal States of the Region (Bahrain, I.R. Iran, Iraq, Kuwait, Oman, Qatar, Saudi Arabia and the United Arab Emirates) to coordinate a common action to protect the semi-enclosed sea surrounded by them. One of its objectives was the knowledge of biodiversity of the marine system. In this framework, the present guide provides a useful tool for identifying more common marine fishes inhabiting the Persian Gulf and the Oman Sea. Like in other otolith catalogues, we show the mesial and lateral view of otolith, but moreover we would like to emphasize that this book contain real fish images, which can be used as an atlas for the region, and graphics of otolith contour based on wavelet's techniques (see material and methods) constituting a relevance change in the structure and format of before catalogues. Finally, we want to highlight that this book is a result of 6 years of labour collecting fish samples and more than 1 year collaborating of scientists from Iran and Spain to give order and writting to this manuscript. We hope it will be a present to otolithologists around the world, being the first step to have a much complete atlas of the Asiatic region.

Material and Methods

Collecting and sampling

Otoliths were collected from 120 species, belonging to 42 families and 7 orders of Teleostean fishes inhabiting the Perisan Gulf and Oman Sea (see data in tables pages 8-10). All specimens were collected from local fisheries markets of Bandar Abbas, Chabahar, Asaluye, Lenge and Dayyer ports. The specimens were fishing using nets, cages (Gargur) or gun for large specimen. All the species belong to the particular collection of Mrs. Zahra Sadighzadeh.

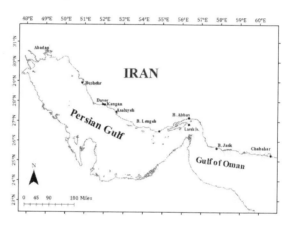

Geographical localitation of studied area and ports where samples were collected

A wooden dhow with cages (letft) and vessel used to fishing by nets or cages (right)

Local fishermen working

Fish sale in the local markets

The taxonomical order used was based on Nelson (2006) and the species were named following the criteria of the online fish catalogue of the California Academy of Science (Eschmeyer et al., 1998). Images of the fish specimen were taken with a digital camera (8 megapixel) at day-light. The best image was selected to illustrate in the present Atlas. For each specimen, the biometry was done, including total length, standard length and total weight.

The left otolith of each pair was used in the figures and graphical illustration of contour. Pictures of the smaller otoliths were taken with a digital camera under a binocular microscope, with the most convenient magnification in each case. Whilst, larger otoliths were photographed by a digital camera (Canon 450D with lense 24-105mm). In both cases, the image was taken of the internal side (medial or proximal) of the otolith as this side presents the sulcus acusticus (a groove along the surface of the sagitta). To obtain a good representation of the sagitta contour, the image must be well contrasted with a homogeneous black background. The otoliths were always represented with the respective dorsal margin to the top of the image and anterior (rostral) region to the right. Terminology and morphological features used in this guide follows Tuset et al. (2008). Finally, otolith length (OL in mm), height (OH in mm), area (OA in mm2) and perimeter (OP in mm) were also measured (see data in tables pages 52-54).

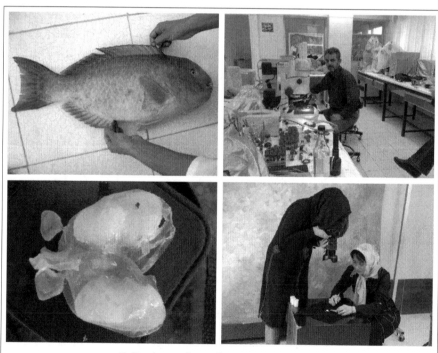

Collecting and sampling in the laboratory

Otolith contour

To provide more information regarding to otolith morphology of each species, we present a graphic illustration of outline otolith using wavelets. It is based on expanding the contour into a family of functions obtained as the dilations and translations of a unique function known as a mother wavelet (Mallat 1991). These functions describe both in space and wave number, the most prominent features of the curve. The signal of wavelets has different amplitudes, hence small wave numbers are associated with a variation in detail of contour while large wave numbers are associated with smoothly contour (Parisi-Baradad et al. 2005, 2010). The otolith contour (512 points) was obtained using a variation of the algorithm proposed in the monograph Morphometrics with R (Claude 2008). Also available in Momocs Package (Bonhomme et al. 2012). Algorithm runs in a clockwise rotation and it has an automatic initial selection point. The initial point was computed as the maximum distance between the centroid and the right part of the otolith. Recently, Sadighzadeh et al. (in press) have demonstrated that wavelet 5 provide the best representation of otolith contour.

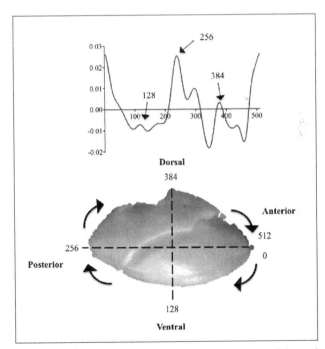

Graphical illustration of wavelet meaning used for the detection of the otolith contour

Checklist of species

Order	Family	Species	Page
Clupeiformes	Chirocentridae	*Chirocentrus dorab* (Forsskål, 1775)	12
		Chirocentrus nudus Swainson, 1839	12
	Clupeidae	*Anodontostoma chacunda* (Hamilton, 1822)	12
		Dussumieria acuta Valenciennes, 1847	13
		Herklostichthys lossei Wongratana, 1983	13
	Pristigasteridae	*Ilisha megaloptera* (Swainson, 1839)	13
	Clupeidae	*Nematolosa nasus* (Bloch, 1795)	14
		Sardinella gibbosa (Bleeker, 1849)	14
Gonorynchiformes	Chanidae	*Chanos chanos* (Forsskål, 1775)	14
Aulopiformes	Synodontidae	*Saurida tumbil* (Bloch, 1795)	15
Beloniformes	Belonidae	*Tylosurus crocodilus* (Péron and Lesueur, 1821)	15
Scorpaeniformes	Platycephalidae	*Platycephalus indicus* (Linnaeus, 1758)	15
Perciformes	Serranidae	*Cephalopholis formosa* (Shaw, 1812)	16
		Cephalopholis hemistiktos (Rüppell, 1830)	16
		Epinephelus bleekeri (Vaillant, 1878)	16
		Epinephelus caeruleopunctatus (Bloch, 1790)	17
		Epinephelus chlorostigma (Valenciennes, 1828)	17
		Epinephelus coioides (Hamilton, 1822)	17
		Epinephelus diacanthus (Valenciennes, 1828)	18
		Epinephelus epistictus (Temminck & Schlegel, 1842)	18
		Epinephelus latifasciatus (Temminck & Schlegel, 1842)	18
	Priacanthidae	*Priacanthus blochii* Bleeker, 1853	19
		Priacanthus tayenus Richardson, 1846	19
	Sillaginidae	*Sillago sihama* (Forsskål, 1775)	19
	Rachycentridae	*Rachycentron canadum* (Linnaeus, 1766)	20
	Carangidae	*Alectis ciliaris* (Bloch, 1787)	20
		Alectis indicus (Rüppell, 1830)	20
		Alepes vari (Cuvier, 1833)	21
		Atropus atropos (Bloch & Schneider, 1801)	21
		Carangoides armatus (Rüppell, 1830)	21
		Carangoides bajad (Forsskål, 1775)	22
		Carangoides chrysophrys (Cuvier, 1833)	22
		Carangoides malabaricus (Bloch & Schneider, 1801)	22
		Caranx sexfasciatus Quoy & Gaimard, 1825	23
		Decapterus tabl Berry, 1968	23
		Gnathanodon speciosus (Forsskål, 1775)	23
		Megalaspis cordyla (Linnaeus, 1758)	24
		Parastromateus niger (Bloch, 1795)	24
		Scomberoides commerson Lacepède, 1801	24
		Scomberoides tol (Cuvier, 1832)	25
		Selar crumenophthalmus (Bloch, 1793)	25
		Selaroides leptolepis (Cuvier, 1833)	25

Checklist of species (Cont.)

Order	Family	Species	Page
		Seriola dumerili (Risso, 1810)	26
		Seriola rivoliana Valenciennes, 1833	26
		Trachinotus mookalee Cuvier, 1832	26
	Menidae	*Mene maculata* (Bloch & Schneider, 1801)	27
	Leiognathidae	*Leiognathus fasciatus* (Lacepède, 1803)	27
	Lutjanidae	*Lutjanus argentimaculatus* (Forsskål, 1775)	27
		Lutjanus bengalensis (Bloch, 1790)	28
		Lutjanus ehrenbergii (Peters, 1869)	28
		Lutjanus erythropterus Bloch, 1790	28
		Lutjanus fulviflamma (Forsskål, 1775)	29
		Lutjanus johnii (Bloch, 1792)	29
		Lutjanus lemniscatus (Valenciennes, 1828)	29
		Lutjanus lutjanus Bloch, 1790	30
		Lutjanus malabaricus (Bloch & Schneider, 1801)	30
		Lutjanus quinquelineatus (Bloch, 1790)	30
		Lutjanus rivulatus (Cuvier, 1828)	31
		Lutjanus russellii (Bleeker, 1849)	31
		Pinjalo pinjalo (Bleeker, 1850)	31
		Pristipomoides sieboldii (Bleeker, 1855)	32
	Lobotidae	*Lobotes surinamensis* (Bloch, 1790)	32
	Gerreidae	*Gerres longirostris* (Lacepède, 1801)	32
		Gerres filamentosus Cuvier, 1829	33
	Haemulidae	*Diagramma pictum* (Thunberg, 1792)	33
		Plectorhinchus flavomaculatus (Cuvier, 1830)	33
		Plectorhinchus gaterinus (Forsskål, 1775)	34
		Plectorhinchus pictus (Tortonese, 1936)	34
		Plectorhinchus schotaf (Forsskål, 1775)	34
		Pomadasys argenteus (Forsskål, 1775)	35
		Pomadasys kaakan (Cuvier, 1830)	35
	Nemipteridae	*Nemipterus japonicus* (Bloch, 1791)	35
		Nemipterus peronii (Valenciennes, 1830)	36
		Scolopsis ghanam (Forsskål, 1775)	36
		Scolopsis taeniatus (Cuvier, 1830)	36
		Scolopsis vosmeri (Bloch, 1792)	37
	Lethrinidae	*Lethrinus microdon* Valenciennes, 1830	37
		Lethrinus nebulosus (Forsskål, 1775)	37
	Sparidae	*Acanthopagrus berda* (Forsskål, 1775)	38
		Acanthopagrus bifasciatus (Forsskål, 1775)	38
		Acanthopagrus latus (Houttuyn, 1782)	38
		Argyrops spinifer (Forsskål, 1775)	39
		Rhabdosargus haffara (Forsskål, 1775)	39
	Polynemidae	*Eleutheronema tetradactylum* (Shaw, 1804)	39
	Sciaenidae	*Otolithes ruber* (Bloch & Schneider, 1801)	40
		Pennahia macrophthalmus (Bloch, 1793)	40
	Mullidae	*Parupeneus rubescens* (Lacepède, 1801)	40
		Upeneus sulphureus Cuvier, 1829	41

Checklist of species (Cont.)

Order	Family	Species	Page
		Upeneus tragula Richardson, 1846	41
	Monodactylidae	*Monodactylus argenteus* (Linnaeus, 1758)	41
	Drepaneidae	*Drepane longimana* (Bloch & Schneider, 1801)	42
		Drepane punctata (Linnaeus, 1758)	42
	Chaetodontidae	*Heniochus acuminatus* (Linnaeus, 1758)	42
	Pomacanthidae	*Pomacanthus maculosus* (Forsskål, 1775)	43
	Terapontidae	*Terapon jarbua* (Forsskål, 1775)	43
		Terapon theraps Cuvier, 1829	43
	Labridae	*Cheilinus lunulatus* (Forsskål, 1775)	44
	Scaridae	*Scarus ghobban* Forsskål, 1775	44
		Scarus persicus Randall & Bruce, 1983	44
	Ephippidae	*Ephippus orbis* (Bloch, 1787)	45
		Platax orbicularis (Forsskål, 1775)	45
	Siganidae	*Siganus javus* (Linnaeus, 1766)	45
		Siganus sutor (Valenciennes, 1828)	46
	Acanthuridae	*Acanthurus sohal* (Forsskål, 1775)	46
	Sphyraenidae	*Sphyraena forsteri* Cuvier, 1829	46
		Sphyraena jello Cuvier, 1829	47
		Sphyraena obtusata Cuvier, 1829	47
		Sphyraena putnamae Jordan & Seale, 1905	47
	Trichiuridae	*Trichiurus lepturus* Linnaeus, 1758	48
	Scombridae	*Auxis thazard thazard* (Lacepède, 1800)	48
		Euthynus affinis (Cantor, 1849)	48
		Rastrelliger kanagurta (Cuvier, 1816)	49
		Scomberomorus commerson (Lacepède, 1800)	49
		Scomberomorus guttatus (Bloch & Schneider, 1801)	49
		Thunnus tonggol (Bleeker, 1851)	50
	Stromateidae	*Pampus argenteus* (Euphrasen, 1788)	50
Pleuronectiformes	Psettodidae	*Psettodes erumei* (Bloch & Schneider, 1801)	51
	Paralichthyidae	*Pseudorhombus elevatus* Ogilby, 1912	51
	Cynoglossidae	*Cynoglossus bilineatus* (Lacepède, 1802)	51

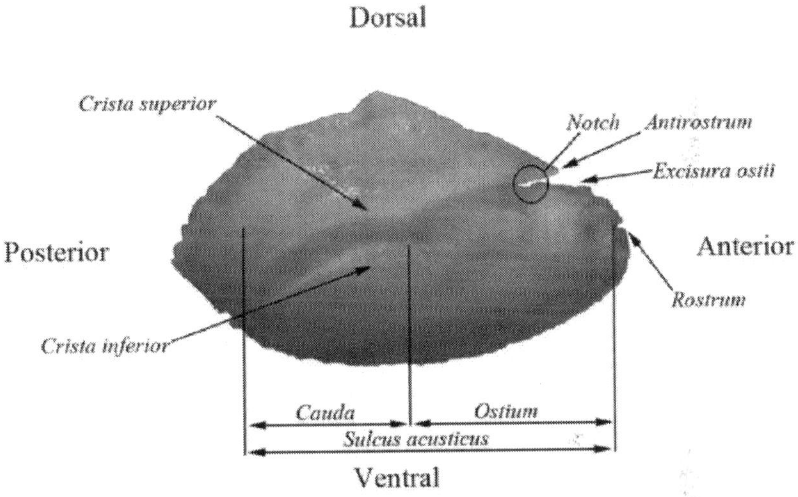
Dorsal
Crista superior
Notch
Antirostrum
Excisura ostii
Posterior
Anterior
Rostrum
Crista inferior
Cauda
Ostium
Sulcus acusticus
Ventral

Morphological descriptions

Family Chirocentridae

Chirocentrus dorab (Forsskål, 1775)

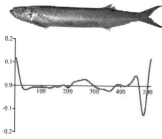

Shape: elliptic. Margins: dorsal margin lobed to anterior with a hump anteriorly, ventral margin crenate. Sulcus acusticus: heterosulcoid, ostial, median. Ostium: funnellike. Cauda: tubular, straight ending far from the posterior margin. Anterior region: peaked; rostrum peaked; antirostrum short, very broad, pointed; excisura wide with an acute deep notch. Posterior region: round.

Chirocentrus nudus Swainson, 1839

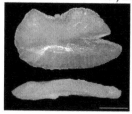
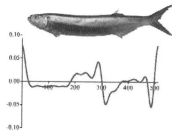

Shape: elliptic-rectangular. Margins: dorsal margin with a fossa in the middle, ventral margin almost entire. Sulcus acusticus: heterosulcoid, ostial, median. Ostium: funnellike. Cauda: tubular, slighly curved ending far from the posterior margin. Anterior region: peaked; rostrum peaked; antirostrum short, very broad, pointed; excisura narrow with an acute shallow notch. Posterior region: round

Family Pristigasteridae

Ilisha megaloptera (Swainson, 1839)

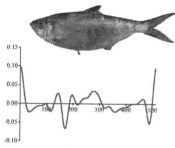

Shape: elliptic-lanceolate. Margins: dorsal margin crenated, ventral margin with dentation in the middle. Sulcus acusticus: heterosulcoid, ostial, median. Ostium: funnellike. Cauda: round-oval, straight ending far from the posterior margin. Anterior region: peaked; rostrum broad, long, pointed and acute; antirostrum broad, short, pointed; excisura wide with an acute deep notch. Posterior region: round to oblique.

Family Clupeidae

Anodostoma chacunda (Hamilton, 1822)

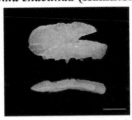
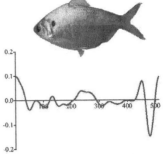

Shape: elliptic-rectangular. Margins: dorsal margin crenate in the middle, ventral margin with irregular dentation. Sulcus acusticus: heterosulcoid, ostial, median. Ostium: funnel-like. Cauda: round-oval, straight ending far from the posterior region. Anterior region: peaked; rostrum broad, medium, pointed; antirostrum long, broad, pointed; excisura wide with an acute deep notch. Posterior region: round.

Herklostichthys lossei Wongratana, 1983

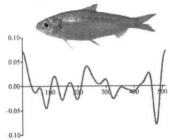

Shape: elliptic. Margins: dorsal margin sinuate, ventral margin dentate. Sulcus acusticus: heterosulcoid, ostial, median. Ostium: funnel-like. Cauda round-oval, straight, ending far from the posterior margin. Anterior region: blunt; rostrum broad, short, pointed; antirostrum short, broad, pointed; excisura wide with an acute shallow notch. Posterior region: round.

Nematolosa nasus (Bloch, 1795)

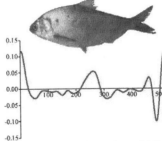

Shape: elliptic-lanceolate. Margins: dorsal margine sinuated in the middle ventral margin serrated. Sulcus acusticus: heterosulcoid, ostial, median. Ostium: funnel-like. Cauda: tubular, straight ending far from the posterior margin. Anterior region: peaked; rostrum broad long pointed; antirostrum short broad pointed; excisura wide with an acute medium notch. Posterior region: round to oblique.

CLUPEIFORMES, GONORYNCHIFORMES, AULOPIFORMES

Sardinella gibbosa (Bleeker, 1849)

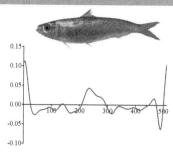

Shape: elliptic-lanceolate. Margins: dorsal margin sinuate, ventral margin serrate. Sulcus acusticus: heterosulcoid, ostial, median. Ostium: funnel-like. Cauda: round-oval, straight ending far from the posterior margin. Anterior region: peaked; rostrum broad long pointed; antirostrum short broad pointed; excisura wide with an acute shallow notch. Posterior region: round.

Family Chanidae

Chanos chanos (Forsskål, 1775)

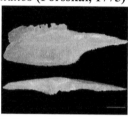
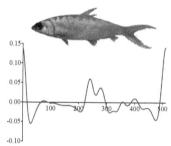

Shape: elliptic-pyriform. Margins: ventral margin crenate. Sulcus acusticus: heterosulcoid, ostial, supramedian. Ostium: tubular. Cauda: tubular, slightly curved ending very close to the posterior margin. Anterior region: lanceolated; rostrum lanceolated; antirostrum absent; excisura without notch. Posterior region: double peaked

Family Synodontidae

Saurida tumbil (Bloch, 1795)

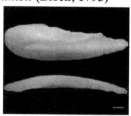
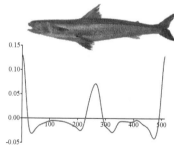

Shape: spindle-shaped. Margins: entire to sinuate. Sulcus acusticus: heterosulcoid, ostial, median. Ostium: funnel-like. Cauda: tubular, straight ending far from the posterior margin. Anterior region: lanceolated; rostrum pointed, long, blunt; antirostrum poorly defined; excisura wide without notch. Posterior region: round.

Family Belonidae

Tylosurus crocodilus **(Péron and Lesueur, 1821)**

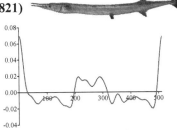

Shape: elliptic. Margins: irregular dorsal margin, ventral margin slightly serrate to irregular. Sulcus acusticus: heterosulcoid, ostial, median. Ostium: funnel-like. Cauda: tubular, straight ending close to the posterior margin. Anterior region: peaked; rostrum peaked; antirostrum absent; excisura without notch. Posterior region: round.

Family Hemirhamphidae

Hemiramphus archipelagicus **Collette & Parin, 1978**

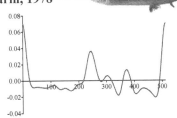

Shape: elliptic. Margins: one dentation in the center of dorsal margin, ventral slightly sinuate. Sulcus acusticus: heterosulcoid, ostial, median. Ostium: funnel-like. Cauda: tubular, straight ending close to the posterior margin. Anterior region: peaked; rostrum peaked, short; antirostrum poorly defined; excisura wide with shallow notch. Posterior region: round-oblique.

Family Platycephalidae

Platycephalus indicus **(Linnaeus, 1758)**

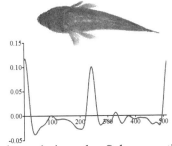

Shape: triangular-lanceolated. Margins: dorsal margin irregular. Sulcus acusticus: heterosulcoid, ostial, median. Ostium: elliptic. Cauda: tubular, slightly curved ending close to the ventral margin. Anterior region: peaked; rostrum broad, long, pointed; antirostrum poorly defined; excisura wide with shallow notch. Posterior region: peaked.

Family Serranidae

Cephalopholis formosa (Shaw, 1812)

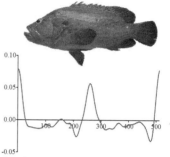

Shape: elliptic-oblong. Margins: dorsal margin sinuate. Sulcus acusticus: heterosulcoid, ostial, median. Ostium: funnel like. Cauda: tubular, strongly curved ending very close to the ventral margin. Anterior region: peaked; rostrum broad, medium, pointed; antirostrum poorly defined; excisura wide with a shallow notch. Posterior region: angled and lobulated.

Cephalopholis hemistiktos (Rüppell, 1830)

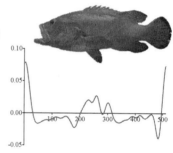

Shape: oblong. Margins: sinuate. Sulcus acusticus: heterosulcoid, ostial, median. Ostium: funnel like. Cauda: tubular, strongly curved ending close to the ventral margin. Anterior region: peaked; rostrum broad, medium, pointed; antirostrum poorly defined; excisura wide with a shallow notch. Posterior region: irregular.

Epinephelus bleekeri (Vaillant, 1878)

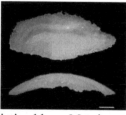
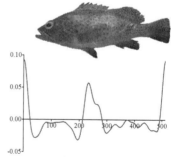

Shape: elliptic-oblong. Margins: anterior dorsal margin irregular, ventral margin sinuate. Sulcus acusticus: heterosulcoid, ostial, median. Ostium: funnel like. Cauda: tubular, strongly curved, ending close to the ventral margin. Anterior region: peaked; rostrum broad, medium, pointed; antirostrum poorly defined; excisura wide with a shallow notch. Posterior region: oblique.

Epinephelus caeruleopunctatus (Bloch, 1790)

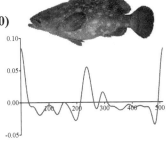

Shape: elliptic-oblong. Margins: dorsal margin crenate. Sulcus acusticus: heterosulcoid, ostial, median. Ostium: funnel like. Cauda: tubular, strongly curved ending very close to the ventral margin. Anterior region: peaked; rostrum broad, medium, pointed; antirostrum poorly defined; excisura wide with shallow notch. Posterior region: oblique.

Epinephelus chlorostigma (Valenciennes, 1828)

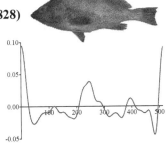

Shape: elliptic-oblong. Margins: dorsal margin irregular to sinuate, ventral margin crenate to sinuate. Sulcus acusticus: heterosulcoid ostial median. Ostium: funnel like. Cauda: tubular, strongly curved ending very close to the ventral margin. Anterior region: peaked; rostrum broad, medium, pointed; antirostrum poorly defined; excisura wide with a shallow notch. Posterior region: oblique.

Epinephelus coioides (Hamilton, 1822)

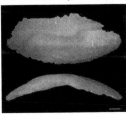
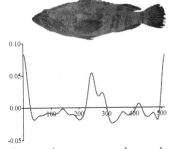

Shape: oblong. Margins: dorsal margin irregular to sinuate, ventral margin crenate. Sulcus acusticus: heterosulcoid ostial median. Ostium: rectangular. Cauda: tubular, strongly curved ending very close to the ventral margin. Anterior region: peaked; rostrum broad, medium, pointed; antirostrum poorly defined; excisura wide with a shallow notch. Posterior region: oblique-irregular.

PERCIFORMES

Epinephelus diacanthus (Valenciennes, 1828)

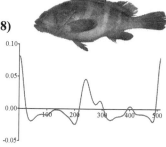

Shape: elliptic-oblong. Margins: crenate. Sulcus acusticus: heterosulcoid, ostial, median. Ostium: funnel-like. Cauda: tubular, strongly curved ending close to the ventral margin. Anterior region: peaked; rostrum broad, medium ,pointed; antirostrum poorly defined; excisura wide with a shallow notch. Posterior region: oblique.

Epinephelus epistictus (Temminck & Schlegel, 1842)

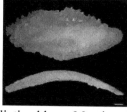
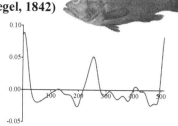

Shape: elliptic-oblong. Margins: dorsal margin serrate to irregular, ventral margin sinuate. Sulcus acusticus: heterosulcoid, ostial, median. Ostium: funnel-like. Cauda: tubular, strongly curved ending close to the ventral margin. Anterior region: peaked; rostrum broad, medium, pointed; antirostrum absent; excisura without notch. Posterior region: irregular.

Epinephelus latifasciatus (Temminck & Schlegel, 1842)

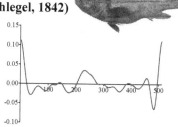

Shape: elliptic. Margins: irregular. Sulcus acusticus: heterosulcoid, ostial, median. Ostium: funnel-like. Cauda: tubular, strongly curved ending close to the ventral margin. Anterior region: peaked; rostrum, broad, long, pointed; antirostrum short, broad, pointed; excisura wide with an acute shallow notch. Posterior region: round-dentate.

Family Priacanthidae

Priacanthus blochii Bleeker, 1853

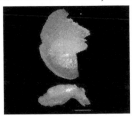
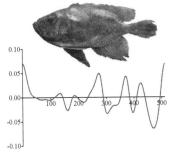

Shape: tall, moon-shape. Margins: crenate dorsal margin. Sulcus acusticus: heterosulcoid, ostial, median. Ostium: oval. Cauda: tubular, slightly curved ending far from the posterior margin. Anterior region: double-peaked; rostrum broad, short, blunt; antirostrum broad, short, blunt; excisura wide with a shallow notch. Posterior region: round.

Priacanthus tayenus Richardson, 1846

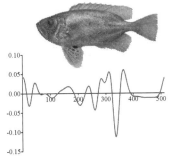

Shape: hour-glass developed ventrally. Margins: dorsal margin sinuate, ventral margin convex and lobed. Sulcus acusticus: heterosulcoid, ostial, supramedian. Ostium: oval. Cauda: tubular slightly curved. Anterior region: double-peaked; rostrum broad, short, pointed; antirostrum broad, short, pointed; excisura wide with an acute deep notch. Posterior region: fish-tail-shaped with two clear identations.

Family Sillaginidae

Sillago sihama (Forsskål, 1775)

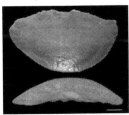
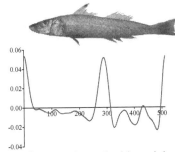

Shape: rhomboidal. Margins: entire. Sulcus acusticus: archaesulcoid, ostial, median. Ostium: tubular undiferentiated of cauda. Cauda: tubular, straight ending very close to the posterior margin. Anterior region: round; rostrum poorly defined; antirostrum poorly defined; excisura wide without notch. Posterior region: peaked.

Family Rachycentridae

Rachycentron canadum (Linnaeus, 1766)

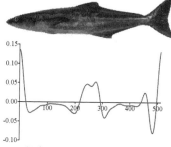

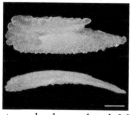

Shape: rectangular-lanceolated. Margins: crenate. Sulcus acusticus: heterosulcoid, ostial, supramedian. Ostium: funnel like. Cauda: oval, straight ending very close to the posterior margin. Anterior region: lanceolated; rostrum narrow, long, pointed; antirostrum broad, short, pointed; excisura wide with an acute shallow notch. Posterior region: flatened and lobulated.

Family Carangidae

Alectis ciliaris (Bloch, 1787)

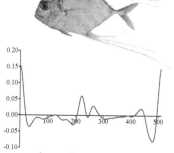

Shape: lanceolated. Margins: crenate. Sulcus acusticus: heterosulcoid, ostiocaudal, median. Ostium: funnel-like. Cauda: elliptic, straight. Anterior region: lanceolated; rostrum narrow, long, pointed; antirostrum short, broad, pointed; excisura wide with an acute shallow notch. Posterior region: irregular; postrostrum, peaked; post-antirostrum, irregular.

Alectis indicus (Rüppell, 1830)

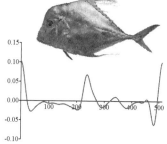

Shape: triangular-elliptic. Margins: dorsal margine sinuate, ventral margin crenate. Sulcus acusticus: heterosulcoid, ostial, median. Ostium: funnel-like. Cauda: tubular, slightly curved ending far from the posterior margin. Anterior region: peaked; rostrum broad, long, pointed; antirostrum short, narrow, pointed; excisura wide with an acute shallow notch. Posterior region: peaked.

Alepes vari (Cuvier, 1833)

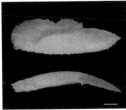 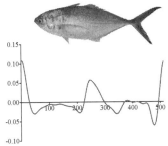

Shape: lanceolated. Margins: dorsal margine with a fossa in the middle, ventral margin sinuate. Sulcus acusticus: heterosulcoid, ostial, median. Ostium: funnel-like. Cauda: tubular, strongly curved ending close to the ventral margin. Anterior region: peaked; rostrum broad, long, pointed; antirostrum short, broad, pointed; excisura narrow with a small shallow notch. Posterior region: round.

Atropus atropos (Bloch & Schneider, 1801)

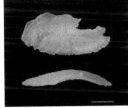 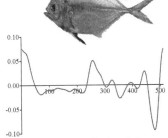

Shape: elliptic-oblong. Margins: ventral margin serrate posteriorly. Sulcus acusticus: heterosulcoid, ostial, median. Ostium: funnel-like. Cauda: tubular, strongly curved ending close to the ventral margin. Anterior region: peaked with a protuberance of ostial colliculum; rostrum broad, long, pointed; antirostrum short, broad, irregular; excisura wide with an acute medium notch. Posterior region: oblique.

Carangoides armatus (Rüppell, 1830)

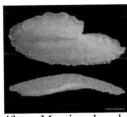 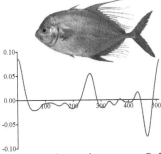

Shape: fusiform. Margins: dorsal margine sinuate, ventral margin crenate. Sulcus acusticus: heterosulcoid, ostial, median. Ostium: funnel-like. Cauda: tubular, strongly curved ending close to the posterior margin. Anterior region: peaked; rostrum broad, long, pointed; antirostrum short, broad, round; excisura wide with an acute medium notch. Posterior region: round.

PERCIFORMES

Carangoides bajad (Forsskål, 1775)

 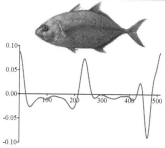

Shape: fusiform. Margins: crenate. Sulcus acusticus: heterosulcoid, ostial, median. Ostium: funnel-like. Cauda: tubular, strongly curved ending close to the ventral margin. Anterior region: peaked with a protuberance of ostial colliculum; rostrum peaked, broad, long, pointed; antirostrum short, broad, pointed; excisura wide with an acute deep notch. Posterior region: oblique.

Carangoides chrysophrys (Cuvier, 1833)

 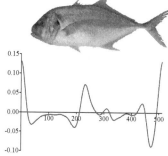

Shape: lanceolated. Margins: dorsal margine sinuate, ventral margin crenate. Sulcus acusticus: heterosulcoid, ostial, median. Ostium: funnel-like. Cauda: tubular, strongly curved ending close to the ventral margin. Anterior region: lanceolated; rostrum narrow, long, pointed; antirostrum short, broad, pointed; excisura wide with an acute notch. Posterior region: oblique.

Carangoides malabaricus (Bloch & Schneider, 1801)

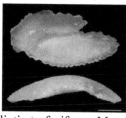 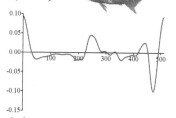

Shape: elliptic to fusiform. Margins: sinuate. Sulcus acusticus: heterosulcoid, ostial, median. Ostium: funnel-like. Cauda: tubular, strongly curved ending close to the ventral margin. Anterior region: peaked; rostrum broad, long, pointed; antirostrum short, broad, pointed; excisura wide with an acute notch. Posterior region: irregular.

PERCIFORMES

Caranx sexfasciatus Quoy & Gaimard, 1825

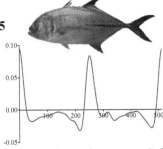

Shape: fusiform. Margins: dorsal margine serrate, ventral margin crenate. Sulcus acusticus: heterosulcoid, ostial, median. Ostium: funnel-like, perforated colliculum. Cauda: tubular, strongly curved ending close to the ventral margin. Anterior region: peaked; rostrum broad, long, pointed; antirostrum absent; excisura without notch. Posterior region: oblique.

Decapterus tabl Berry, 1968

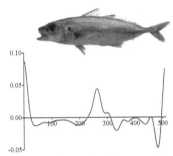

Shape: elliptic. Margins: dorsal margine sinuate, ventral margin almost entire. Sulcus acusticus: heterosulcoid, ostial, median. Ostium: funnel-like. Cauda: tubular, strongly curved ending close to the ventral margin. Anterior region: peaked; rostrum broad, medium, pointed; antirostrum short, broad, blunt; excisura wide with an acute shallow notch. Posterior region: peaked.

Gnathanodon speciosus (Forsskål, 1775)

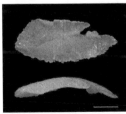
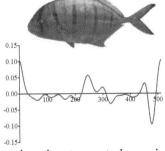

Shape: elliptic- fusiform. Margins: dorsal margine sinuate, ventral margin crenate. Sulcus acusticus: heterosulcoid, ostial, median. Ostium: funnel-like. Cauda: tubular, strongly curved ending close to the posterior margin. Anterior region: peaked with a protuberance of ostial colliculum; rostrum broad, medium, pointed; antirostrum short, narrow, pointed; excisura wide with an acute medium notch. Posterior region: oblique.

PERCIFORMES

Megalaspis cordyla (Linnaeus, 1758)

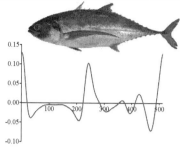

Shape: lanceolated. Margins: sinuate. Sulcus acusticus: heterosulcoid, ostial, median. Ostium: funnel-like. Cauda: tubular, strongly curved ending very close to the ventral margin. Anterior region: lanceolated with dentate protuberances; rostrum narrow, long, pointed; antirostrum very short, broad, pointed upward; excisura wide with an square medium notch. Posterior region: oblique.

Parastromateus niger (Bloch, 1795)

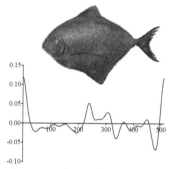

Shape: oblong. Margins: dorsal margine irregular, ventral margin crenate to irregular. Sulcus acusticus: heterosulcoid, ostial, smedian. Ostium: funnel-like. Cauda: tubular, slightly curved ending very close to the ventral margin. Anterior region: peaked; rostrum broad, long, pointed; antirostrum short, broad, pointed; excisura wide with an acute shallow notch. Posterior region: irregular.

Scomberoides commerson Lacepède, 1801

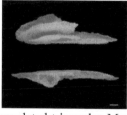
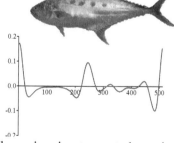

Shape: lanceolated-triangular. Margins: dorsal margine sinuate, ventral margin serrate. Sulcus acusticus: heterosulcoid, ostial, median. Ostium: funnel-like. Cauda: tubular slightly curved ending very close to the ventral margin. Anterior region: lanceolated; rostrum narrow, long, pointed; antirostrum short, broad, pointed; excisura wide with an acute medium notch. Posterior region: round.

PERCIFORMES

Scomberoides tol (Cuvier, 1832)

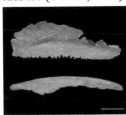 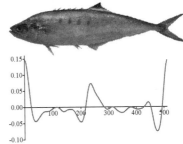

Shape: lanceolated. Margins: dorsal margin iregular, ventral margin serrate to dentate. Sulcus acusticus: heterosulcoid, ostial, median. Ostium: funnel-like. Cauda: tubular, slightly curved ending very close to the posterior margin. Anterior region: lanceolated; rostrum narrow, long, pointed; antirostrum short, broad, pointed; excisura wide with an very acute medium notch. Posterior region: irregular.

Selar crumenophthalmus (Bloch, 1793)

 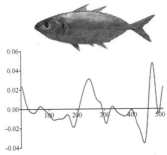

Shape: elliptic. Margins: dorsal margine entire, ventral margin serrate to sinuate. Sulcus acusticus: heterosulcoid, ostial, median. Ostium: oval. Cauda: tubular, markedly curved ending very close to the ventral margin. Anterior region: peaked with one big protuberance; rostrum broad, short, pointed; antirostrum very small; excisura tiny notch. Posterior region: round.

Selaroides leptolepis (Cuvier, 1833)

 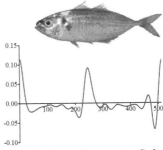

Shape: elliptic. Margins: dorsal margin sinuate, ventral margin crenate. Sulcus acusticus: heterosulcoid, ostial, median. Ostium: funnel-like. Cauda: tubular, strongly curved ending close to the ventral margin. Anterior region: pointed; rostrum long, broad, pointed; antirostrum narrow, short, blunt; excisura wide with an shallow notch. Posterior region: oblique.

PERCIFORMES

Seriola dumerili (Risso, 1810)

 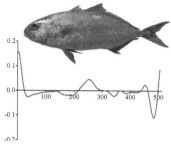

Shape: lanceolated. Margins: dorsal margin lobed, ventral margin serrate to dentate. Sulcus acusticus: heterosulcoid, ostial, median. Ostium: funnel-like. Cauda: tubular, strongly curved ending far from the posterior margin. Anterior region: lanceolated; rostrum narrow, long, pointed; antirostrum short, narrow, pointed; excisura wide with an acute deep notch. Posterior region: angled.

Seriola rivoliana Valenciennes, 1833

 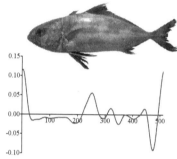

Shape: lanceolated. Margins: dorsal margin lobed, ventral margin serrate to crenate. Sulcus acusticus: heterosulcoid, ostial, median. Ostium: funnel-like. Cauda: tubular, strongly curved ending close to the posterior margin. Anterior region: peaked; rostrum broad, long, pointed; antirostrum short, broad, blunt; excisura wide with an acute shallow notch. Posterior region: oblique.

Trachinotus mookalee Cuvier, 1832

 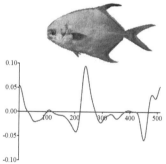

Shape: fusiform. Margins: irregular to crenate. Sulcus acusticus: heterosulcoid, ostial, median. Ostium: funnel-like. Cauda: tubular, slightly curved ending close to the posterior margin. Anterior region: irregular with a protuberance; rostrum long, wide pointed upward; antirostrum short, broad, round; excisura wide with an acute shallow notch. Posterior region: oblique.

Family Menidae
Mene maculata (Bloch & Schneider, 1801)

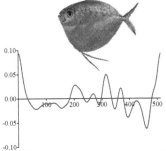

Shape: triangular with mid-dorssl concavity. Margins: dorsal margine irregular posteriorly, ventral margin crenate. Sulcus acusticus: heterosulcoid, ostial, median. Ostium: funnel-like. Cauda: tubular, slighly curved endingvery close to the posterior margin. Anterior region: peaked; rostrum medium, broad, pointed; antirostrum short, broad, blunt; excisura wide with an obtuse shallow notch. Posterior region: flattened.

Family Leiognathidae
Leiognathus fasciatus (Lacepède, 1803)

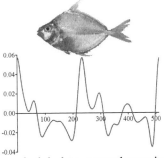

Shape: pentagonal. Margins: anterior-dorsal margin lobulate, ventral margin serrated anteriorly. Sulcus acusticus: heterosulcoid, ostial, median. Ostium: funnel like. Cauda: tubular, slightly curved ending close the posterior margin. Anterior region: peaked; rostrum short, broad, pointed; antirostrum poorly defined; excisura wide with shallow notch. Posterior region: oblique.

Family Lutjanidae
Lutjanus argentimaculatus (Forsskål, 1775)

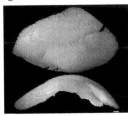
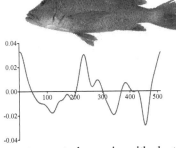

Shape: pentagonal. Margins: dorsal margin crenate, ventral margin with dentations. Sulcus acusticus: heterosulcoid, ostial, median. Ostium: funnel like. Cauda tubular, markedly curved ending close to the posterior ventral margin. Anterior region: peaked to angled; rostrum short, broad, round; antirostrum short, broad, pointed; excisura narrow with a shallow acute notch. Posterior region: oblique-irregular.

Lutjanus bengalensis (Bloch, 1790)

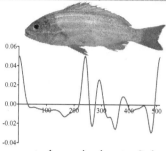

Shape: elliptic. Margins: dorsal margin irregular, ventral margin sinuate. Sulcus acusticus: heterosulcoid, ostial, median. Ostium: rectangular. Cauda tubular, strongly curved and flexed from the middle region ending very close to the ventral margin. Anterior region: peaked to angled; rostrum short, broad, blunt; antirostrum poorly defined; excisura wide without notch. Posterior region: oblique-irregular.

Lutjanus ehrenbergii (Peters, 1869)

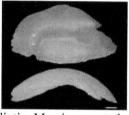
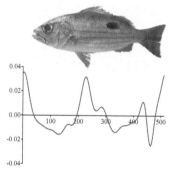

Shape: elliptic. Margins: ventral margin crenate to irregular. Sulcus acusticus: heterosulcoid, ostial, median. Ostium: funnel like. Cauda tubular, strongly curved ending very close to the ventral margin. Anterior region: peaked; rostrum short, broad, pointed; antirostrum short, broad, pointed; excisura wide with a shallow and acute notch. Posterior region: oblique.

Lutjanus erythropterus Bloch, 1790

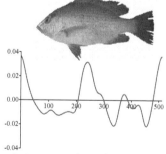

Shape: pentagonal to elliptic, Margins: crenate. Sulcus acusticus: heterosulcoid, ostial, median. Ostium: funnel like. Cauda: tubular, strongly curved and flexed from the middle region ending very close to the ventral margin. Anterior region: peaked to round; rostrum short, broad, pointed to round; antirostrum short, broad, pointed; excisura wide with a shallow or deep and acute notch. Posterior region: oblique.

Lutjanus fulviflamma (Forsskål, 1775)

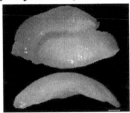 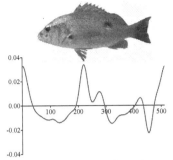

Shape: pentagonal. Margins: margins entire to crenate. Sulcus acusticus: heterosulcoid, ostial, median. Ostium: funnel like. Cauda: tubular, strongly curved and flexed from the middle region ending very close to the ventral margin. Anterior region: peaked; rostrum short, broad, blunt; antirostrum absent or small, narrow, pointed; excisura wide with a deep and acute notch. Posterior region: oblique.

Lutjanus johnii (Bloch, 1792)

 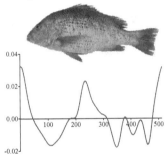

Shape: elliptic. Margins: ventral margins crenate to sinuate. Sulcus acusticus: heterosulcoid, ostial, median. Ostium: funnel like to rectangular. Cauda: tubular, slightly curved ending close to the ventral margin. Anterior region: peaked to round; rostrum short, broad, round; antirostrum poorly defined or small, broad, pointed; excisura wide without or with a small shallow notch. Posterior region: oblique to angled.

Lutjanus lemniscatus (Valenciennes, 1828)

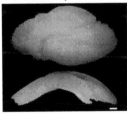 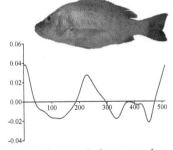

Shape: elliptic. Margins: ventral margins crenate to sinuate. Sulcus acusticus: heterosulcoid, ostial, median. Ostium: funnel-lik. Cauda: tubular, markedly curved flexed from the middle region ending close to the ventral margin. Anterior margin: peaked to round; rostrum short, broad, pointed; antirostrum poorly defined or very small, broad, pointed; excisura wide with a shallow and acute notch. Posterior region: oblique.

Lutjanus lutjanus Bloch, 1790

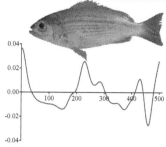

Shape: elliptic to pentagonal. Margins: ventral margin sinuate to crenate. Sulcus acusticus: heterosulcoid, ostial, median. Ostium: funnel-like. Cauda: tubular, strongly curved ending very close to the ventral margin. Anterior margin: peaked; rostrum short, broad, pointed; antirostrum small, broad, pointed; excisura wide with a shallow and acute notch. Posterior region: oblique to angled.

Lutjanus malabaricus (Bloch & Schneider, 1801)

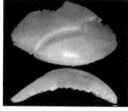
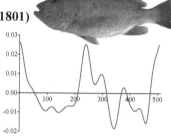

Shape: pentagonal to elliptic. Margins: crenate margins, dorsal margin irregular. Sulcus acusticus: heterosulcoid, ostial, median. Ostium: funnel-like to rectangular. Cauda: tubular, slightly curved ending very close to the ventral margin. Anterior margin: peaked-angled; rostrum short, broad, round; antirostrum poorly defined; excisura wide without notch. Posterior region: oblique to angled.

Lutjanus quinquelineatus (Bloch, 1790)

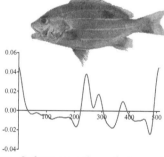

Shape: pentagonal. Margins: sinuate to crenate. Sulcus acusticus: heterosulcoid, ostial, median. Ostium: funnel like. Cauda: tubular, strongly curved and flexed from the middle regio ending close to the ventral margin. Anterior region: peaked-angled; rostrum short, broad, round; antirostrum poorly defined; excisura wide with a shallow notch. Posterior region: oblique.

Lutjanus rivulatus (Cuvier, 1828)

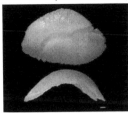
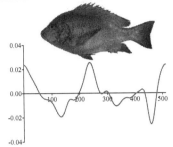

Shape: pentagonal to elliptic. Margins: ventral margins crenate to dentate. Sulcus acusticus: heterosulcoid, ostial, median. Ostium: funnel-like, almost as long as cauda. Cauda: tubular, markedly curved and flexed from the middle region, ending close to the ventral margin. Anterior margin: round; rostrum short, broad, round; antirostrum absent or small, broad, pointed; excisura wide without or with a deep and acute notch. Posterior region: oblique.

Lutjanus russelli (Bleeker, 1849)

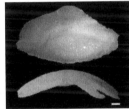
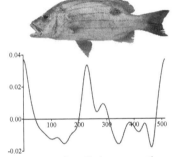

Shape: pentagonal. Margins: dorsal margin serrate to entire. Sulcus acusticus: heterosulcoid, ostial, median. Ostium: funnel-like. Cauda: tubular, strongly curved and flexed from the middle region ending close to the middle of ventral margin. Anterior margin: peaked to blunt; rostrum, short, broad, blunt; antirostrum poorly developed or absent; excisura wide with or without notch. Posterior region: oblique.

Pinjalo pinjalo (Bleeker, 1850)

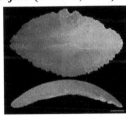
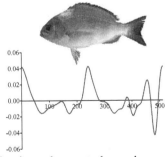

Shape: elliptic to oval. Margins: dorsal margine irregular ventral margin crenate. Sulcus acusticus: heterosulcoid, ostial, median. Ostium: funnel-like. Cauda: tubular. strongly curved ending very close to the posterior margin. Anterior margin: peaked; rostrum short broad round; antirostrum poorly defined; excisura wide with a shallow notch. Posterior region: angled.

Pristipomoides sieboldii (Bleeker, 1855)

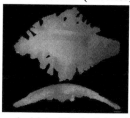 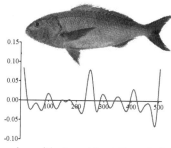

Shape: rhomboidal-elliptic. Margins: very irregular with deep identations in large specimens. Sulcus acusticus: heterosulcoid, ostial, median. Ostium: rectangular. Cauda: tubular, slightly curved ending very close to the posterior margin. Anterior margin: peaked; rostrum long, broad, blunt; antirostrum short, broad, pointed; excisura wide with an acute shallow notch. Posterior region: angled.

Family Lobotidae

Lobotes surinamensis (Bloch, 1790)

 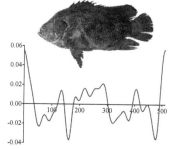

Shape: oval. Margins: irregular with dentations. Sulcus acusticus: heterosulcoid, ostial, median. Ostium: funnel like. Cauda tubular strongly curved ending very close to the posterior margin. Anterior region: peaked with a protuberance; rostrum short, broad, round; antirostrum poorly defined; excisura wide with a shallow notch. Posterior region: round.

Family Gerreidae

Gerres filamentosus Cuvier, 1829

 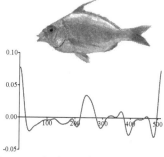

Shape: elliptic-pentagonal. Margins: posterior margin irregular. Sulcus acusticus: heterosulcoid, ostial, median. Ostium: funnel like. Cauda: tubular, strongly curved ending close to the posterior margin. Anterior region: peaked; rostrum medium, broad, peaked; antirostrum short, broad; excisura wide with shallow acute notch. Posterior region: round-irregular.

Gerres longirostris (Lacepède, 1801)

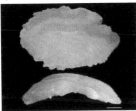
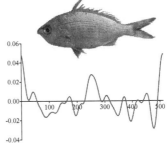

Shape: elliptic. Margins: dorsal margin with a fossa in the middle, ventral margin serrate to dentate. Sulcus acusticus: heterosulcoid, ostial, supramedian. Ostium: funnel like. Cauda: tubular, strongly curved ending close to the posterior margin. Anterior region: peaked; rostrum broad, medium, pointed; antirostrum short, broad, pointed; excisura wide with a shallow notch. Posterior region: round.

Family Haemulidae

Diagramma pictum (Thunberg, 1792)

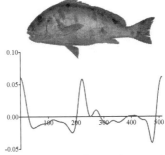

Shape: pentagonal. Margins: dorsal margine dentate, ventral margin serrate. Sulcus acusticus: heterosulcoid, ostial, median. Ostium: rectangular. Cauda: tubular, curled ending close to the ventral margin. Anterior region: round; rostrum medium, broad, round; antirostrum poorly defined; excisura wide with a shallow acute notch. Posterior region: oblique.

Plectorhinchus flavomaculatus (Cuvier, 1830)

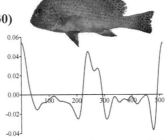

Shape: elliptic. Margins: dorsal margine dentate, ventral margin serrate. Sulcus acusticus: heterosulcoid, ostial, median. Ostium: rectangular. Cauda: tubular, strongly curved ending close to the ventral margin. Anterior region: round; rostrum short, broad, round; antirostrum short, narrow, pointed; excisura wide with an acute shallow notch. Posterior region: oblique.

PERCIFORMES

Plectorhinchus gaterinus (Forsskål, 1775)

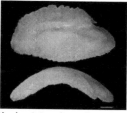
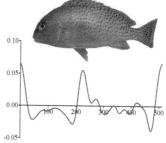

Shape: elliptic. Margins: dorsal margin serrate, posteriorly dentated in ventral-posterior margin. Sulcus acusticus: heterosulcoid, ostial, median. Ostium: rectangular. Cauda: tubular, curled ending close to the ventral margin. Anterior region: peaked; rostrum short, broad, round; antirostrum short, narrow, pointed; excisura wide with an acute shallow notch. Posterior region: oblique.

Plectorhinchus pictus (Tortonese, 1936)

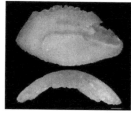
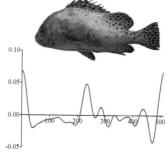

Shape: elliptic. Margins: dorsal margine dentate, ventral margin serrate. Sulcus acusticus: heterosulcoid, ostial, median. Ostium: rectangular. Cauda: tubular, markedly curved ending close to the ventral margin. Anterior region: peaked; rostrum short, broad, round; antirostrum short, narrow, pointed; excisura wide with an acute shallow notch. Posterior region: oblique.

Plectorhinchus schotaf (Forsskål, 1775)

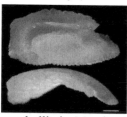
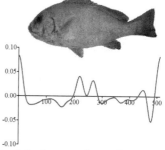

Shape: squared-elliptic. Margins: dorsal margine lobed, ventral margin crenate. Sulcus acusticus: heterosulcoid, ostial, median or supramedian. Ostium: rectangular. Cauda: tubular, curled ending close to the ventral margin. Anterior region: peaked; rostrum short, broad, pointed; antirostrum short, narrow, pointed; excisura wide with an acute shallow notch. Posterior region: flattened.

Pomadasys argenteus (Forsskål, 1775)

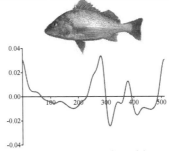

Shape: oval, ventral area very developed. Margins: dorsal margin with apex in the middle ventral margin serrate. Sulcus acusticus: heterosulcoid, ostial, median. Ostium: rectangular. Cauda: tubular, strongly curved ending close to the ventral margin. Anterior region: round; rostrum short, broad, round; antirostrum absent; excisura without notch. Posterior region: angled.

Pomadasys kaakan (Cuvier, 1830)

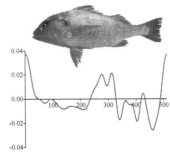

Shape: round-oval, ventral area very developed. Margins: dorsal margin irregular, ventral margin serrate. Sulcus acusticus: heterosulcoid, ostial, median. Ostium: rectangular. Cauda: tubular, strongly curved ending close to the ventral margin. Anterior region: round; rostrum short, broad, round; antirostrum absent; excisura without notch. Posterior region: angled.

Family Nemipteridae

Nemipterus japonicus (Bloch, 1791)

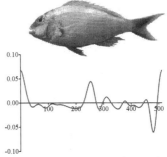

Shape: elliptic-pentagonal. Margins: dorsal margine dentate, ventral margin crenate. Sulcus acusticus: heterosulcoid, ostial, median. Ostium: funnel-like. Cauda: tubular, strongly curved ending close to the ventral margin. Anterior region: peaked; rostrum medium, broad, pointed; antirostrum short, broad, pointed; excisura wide with an acute shallow notch. Posterior region: angled.

Nemipterus peronii (Valenciennes, 1830)

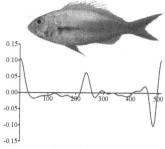

Shape: elliptic-fusiform. Margins: sinuate. Sulcus acusticus: heterosulcoid, ostial, median. Ostium: funnel-like. Cauda: tubular, strongly curved ending close to the ventral margin. Anterior region: peaked; rostrum medium, broad, pointed; antirostrum short, broad, round; excisura wide with an acute deep notch. Posterior region: angled-oblique.

Scolopsis ghanam (Forsskål, 1775)

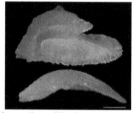
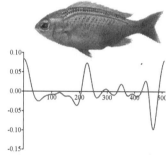

Shape: triangular-elliptic. Margins: dorsal margine with apex in the middle, ventral-margin serrate. Sulcus acusticus: heterosulcoid, ostial, median. Ostium: funnel-like. Cauda: tubular, markedly curved ending close to the ventral margin. Anterior region: peaked; rostrum medium, broad, pointed; antirostrum short, broad, pointed; excisura wide with an acute deep notch. Posterior region: angled-oblique.

Scolopsis taeniatus (Cuvier, 1830)

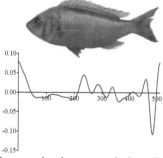

Shape: rectangular-fusiform. Margins: a distinct projection posteriodorsally, ventral margine sinuate. Sulcus acusticus: heterosulcoid, ostial, median. Ostium: funnel-like. Cauda: tubular, strongly curved ending very close to the posterior margin. Anterior region: peaked; rostrum medium, broad, blunt; antirostrum long, narrow, pointed; excisura narrow with an acute deep notch. Posterior region: oblique-irregular.

Scolopsis vosmeri (Bloch, 1792)

 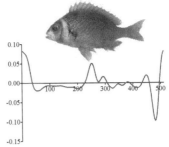

Shape: rectangular-fusiform. Margins: dorsal margine lobed, ventral margin sinuate. Sulcus acusticus: heterosulcoid, ostial, median. Ostium: funnel-like. Cauda: tubular, strongly curved ending close to the posterior margin. Anterior region: peaked; rostrum medium, broad, blunt; antirostrum short broad, round; excisura wide with an acute shallow notch. Posterior region: angled-oblique with notch.

Family Lethrinidae

Lethrinus microdon Valenciennes, 1830

 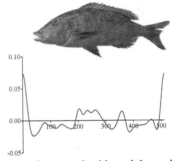

Shape: elliptic. Margins: sinuate. Sulcus acusticus: heterosulcoid, ostial, median. Ostium: rectangular. Cauda: tubular markedly curved ending ending close to the ventral margin . Anterior region: peaked; rostrum short, broad, pointed; antirostrum poorly defined; excisura wide with shallow notch. Posterior region: round-irregular.

Lethrinus nebulosus (Forsskål, 1775)

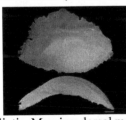 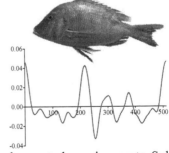

Shape: elliptic. Margins: dorsal margine irregular, ventral margin crenate. Sulcus acusticus: heterosulcoid, ostial, median. Ostium: rectangular. Cauda: tubular, strongly curved ending close to the ventral margin. Anterior region: round; rostrum short, broad, blunt; antirostrum absent; excisura without notch. Posterior region: angled-oblique with dentation.

PERCIFORMES

Family Sparidae

Acanthopagrus berda (Forsskål, 1775)

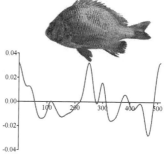

Shape: elliptic. Margins: sinuate with small dentations. Sulcus acusticus: heterosulcoid, ostial, supramedian. Ostium: funnel-like. Cauda: tubular, strongly curved ending far from the ventral margin. Anterior region: round; rostrum short, broad, round; antirostrum poorly defined; excisura without notch. Posterior region: round.

Acanthopagrus bifasciatus (Forsskål, 1775)

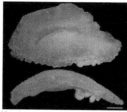
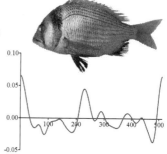

Shape: elliptic. Margins: sinuate with small dentations. Sulcus acusticus: heterosulcoid, ostial, supramedian. Ostium: funnel-like. Cauda: tubular, strongly curved ending far from the ventral margin. Anterior region: peaked; rostrum medium, broad, pointed; antirostrum poorly defined; excisura wide, dentate with a very shallow notch. Posterior region: oblique.

Acanthopagrus latus (Houttuyn, 1782)

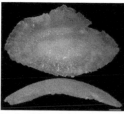
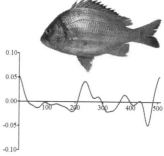

Shape: elliptic. Margins: sinuate with small dentations in the middle of dorsal region. Sulcus acusticus: heterosulcoid, ostial, supramedian. Ostium: funnel-like. Cauda: tubular, strongly curved ending close to the ventral margin. Anterior region: pointed to angled; rostrum medium, broad, pointed; antirostrum short, broad, pointed; excisura wide with an acute shallow notch. Posterior region: oblique.

Argyrops spinifer (Forsskål, 1775)

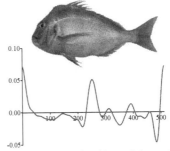

Shape: elliptic. Margins: dentate. Sulcus acusticus: heterosulcoid, ostial, median. Ostium: funnel-like. Cauda: tubular, slightly curved curved ending far from the posterior margin. Anterior region: peaked; rostrum medium, broad, pointed; antirostrum short, broad, pointed; excisura wide with an acute shallow notch. Posterior region: angled with dentations.

Rhabdosargus haffara (Forsskål, 1775)

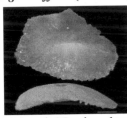
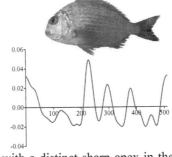

Shape: pentagonal. Margins: dorsal margin with a distinct sharp apex in the middle, ventral margin crenate. Sulcus acusticus: heterosulcoid, ostial, median. Ostium: funnel-like. Cauda: tubular, strongly curved ending far from the ventral margin. Anterior region: round; rostrum short, broad, round; antirostrum short, broad, round; excisura wide with an acute shallow notch. Posterior region: oblique.

FamilyPolynemidae

Eleutheronema tetradactylum (Shaw, 1804)

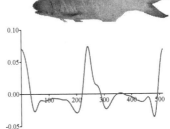

Shape: fusiform. Margins: crenate. Sulcus acusticus: heterosulcoid, ostial, median. Ostium: funnel-like. Cauda: tubular, strongly curved ending very close to the ventral margin. Anterior region: peaked; rostrum medium, broad, round; antirostrum short, broad, pointed; excisura wide with an acute shallow notch. Posterior region: oblique.

PERCIFORMES

Family Scianidae

Otolithes ruber (Bloch & Schneider, 1801)

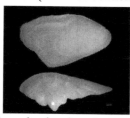 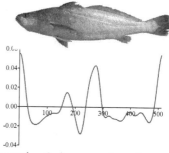

Shape: triangular developed ventrally. Margins: entire. Sulcus acusticus: heterosulcoid, pseudo-ostial, supramedian. Ostium: lateral very developed. Cauda tubular, curled ending very clsoe to the ventral margin. Anterior region: peaked; rostrum medium, wide, round; antirostrum absent; excisura absent. Posterior region: flattened.

Pennahia macrophthalmus (Bloch, 1793)

 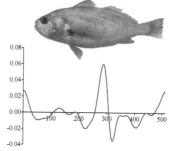

Shape: oval-triangular. Margins: entire. Sulcus acusticus: heterosulcoid, pseudo-ostial, supramedian. Ostium: lateral very developed. Cauda tubular, straight ending far from the posterior margin. Anterior region: round; rostrum medium, wide, round; antirostrum absent; excisura absent. Posterior region: oblique.

Family Mullidae

Parupeneus rubescens (Lacepède, 1801)

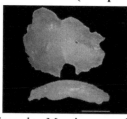 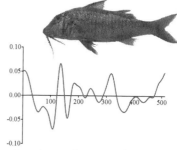

Shape: triangular. Margins: apex in posterior dorsal margin, ventral margin with deep identations. Sulcus acusticus: heterosulcoid, ostial, median. Ostium: funnel-like. Cauda: tubular, strongly curved ending very close to the posterior margin. Anterior region: blunt; rostrum medium, wide, irregular; antirostrum poorly defined; excisura without notch. Posterior region: irregular.

Upeneus sulphureus Cuvier, 1829

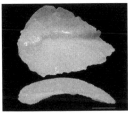
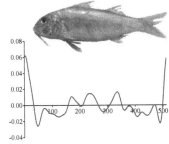

Shape: oval. Margins: sinuate to crenate. Sulcus acusticus: heterosulcoid, ostial, median. Ostium: funnel-like. Cauda: tubular, strongly curved ending very close to the posterior margin. Anterior region: peaked; rostrum medium, broad, pointed; antirostrum poorly defined; excisura narrow with an acute shallow notch. Posterior region: round.

Upeneus tragula Richardson, 1846

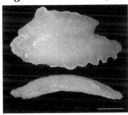
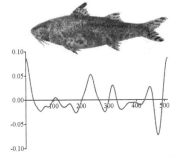

Shape: elliptic. Margins: apex in posterior dorsal margin, ventral margin dentate to lobed. Sulcus acusticus: heterosulcoid, ostial, median. Ostium: funnel-like. Cauda: tubular, slightly curved ending close to the posterior margin. Anterior region: peaked; rostrum medium, broad, pointed; antirostrum short, broad, pointed; excisura wide with an acute shallow notch. Posterior region: oblique-angled.

Family Monodactylidae

Monodactylus argenteus (Linnaeus, 1758)

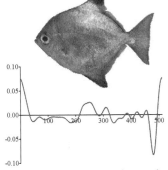

Shape: elliptic-pentagonal. Margins: dorsal margine dentate, ventral margin crenate. Sulcus acusticus: heterosulcoid, ostial, median. Ostium: funnel-like. Cauda: tubular, strongly curved ending close to the posterior margin. Anterior region: peaked; rostrum medium, broad, pointed; antirostrum short, broad, round; excisura narrow with an acute shallow notch. Posterior region: round.

Family Drepaneidae

Drepane longimana (Bloch & Schneider, 1801)

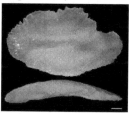
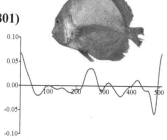

Shape: elliptic-pentagonal. Margins: dorsal margin serrate, ventral margin sinuate. Sulcus acusticus: heterosulcoid, ostial, median. Ostium: rectangular. Cauda: tubular, strongly curved ending very close to the ventral margin. Anterior region: peaked; rostrum medium, broad, round; antirostrum poorly defined; excisura narrow with a shallow notch. Posterior region: round.

Drepane punctata (Linnaeus, 1758)

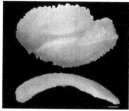
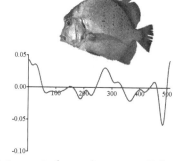

Shape: elliptic. Margins: dorsal margin dentate, ventral margin serrate. Sulcus acusticus: heterosulcoid, ostial, median. Ostium: rectangular. Cauda: tubular, strongly curved ending very close to the ventral margin. Anterior region: peaked; rostrum medium, broad, blunt; antirostrum poorly defined; excisura narrow with a shallow notch. Posterior region: round.

Family Chaetodontidae

Heniochus acuminatus (Linnaeus, 1758)

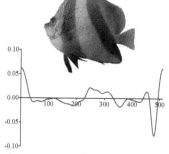

Shape: pentagonal. Margins: dorsal margin sinuate, ventral margin serrate. Sulcus acusticus: heterosulcoid, ostial, median. Ostium: funnel-like. Cauda: tubular, strongly curved ending close to the posterior margin. Anterior region: peaked; rostrum medium, broad, blunt; antirostrum short, broad, round; excisura narrow with an acute shallow notch. Posterior region: oblique.

Family Pomacanthidae

Pomacanthus maculosus (Forsskål, 1775)

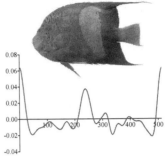

Shape: elliptic. Margins: crenate anteriorly with a notch in dorsal posterior margin. Sulcus acusticus: heterosulcoid ostial supramedian. Ostium: funnel-like. Cauda: tubular, strongly curved close to the ventral margin. Anterior region: peaked; rostrum medium, broad, pointed; antirostrum absent; excisura narrow without notch. Posterior region: round.

Family Terapontidae

Terapon jarbua (Forsskål, 1775)

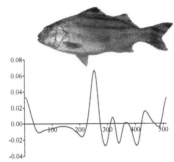

Shape: elliptic. Margins: dorsal margin dorsal margin dentate, ventral margin sinuate to serrate. Sulcus acusticus: heterosulcoid, ostial, median. Ostium: funnel. Cauda: tubular, markaely curved ending far from the posterior margin. Anterior region: round-angled; rostrum very short, broad, round; antirostrum poorly defined; excisura narrow without notch. Posterior region: peaked-pointed.

Terapon theraps Cuvier, 1829

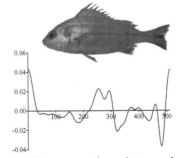

Shape: elliptic. Margins: ventral margin serrate. Sulcus acusticus: heterosulcoid, ostial, median. Ostium: rectangular. Cauda: tubular, markedly curved ending close to the ventral margin. Anterior region: peaked; rostrum short, broad, blunt; antirostrum short, broad, pointed; excisura wide with a shallow and acute notch. Posterior region: round.

PERCIFORMES

Family Labridae
Cheilinus lunulatus (Forsskål, 1775)

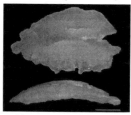
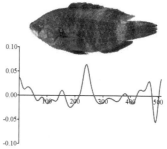

Shape: cuneiform. Margins: irregular. Sulcus acusticus: heterosulcoid, ostial, median. Ostium: funnel like. Cauda: elliptic, ending very close to the posterior margin. Anterior region: double peaked; rostrum short, broad, pointed; antirostrum medium, very broad, pointed; excisura wide with an acute notch. Posterior region: pointed.

Family Scaridae
Scarus ghobban Forsskål, 1775

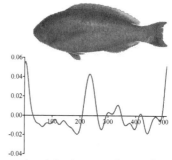

Shape: cuneiform-elliptic. Margins: dorsal margine lobed, ventral margin serrate to dentate. Sulcus acusticus: heterosulcoid, ostio, median or supramedian. Ostium: funnel like. Cauda: elliptic, ending close to the posterior margin. Anterior region: peaked; rostrum short, broad, pointed; antirostrum poorly defined; excisura without notch. Posterior region: angled-pointed.

Scarus persicus Randall & Bruce, 1983

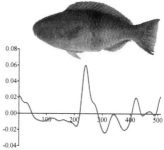

Shape: cuneiform-oval. Margins: dorsal anterior margin with an apex, ventral margin sinuate. Sulcus acusticus: heterosulcoid, ostial, median. Ostium: funnel like. Cauda: elliptic, ending close the posterior margin. Anterior region: round; rostrum short, broad, blunt; antirostrum short, broad, round; excisura narrow with a shallow squared notch. Posterior region: angled-pointed.

Family Ephippidae

Ephippus orbis (Bloch, 1787)

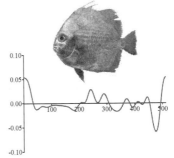

Shape: pentagonal. Margins: dorsal margin irregular, ventral margin sinuate. Sulcus acusticus: heterosulcoid, ostial, median. Ostium: funnel like. Cauda: tubular strongly curved ending close to the posterior margin. Anterior region: blunt; rostrum medium, broad, blunt; antirostrum short, broad, round; excisura wide with an acute shallow notch. Posterior region: oblique.

Platax orbicularis (Forsskål, 1775)

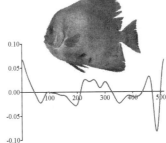

Shape: rectangular-elliptic. Margins: dorsal margin slightly lobbed, ventral margin crenate. Sulcus acusticus: heterosulcoid, ostial, median. Ostium: funnel like. Cauda: elliptic, very ending close to the posterior margin. Anterior region: peaked; rostrum medium, broad, blunt; antirostrum short, broad, pointed; excisura wide with an acute deep notch. Posterior region: round.

Family Siganidae

Siganus javus (Linnaeus, 1766)

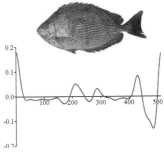

Shape: elliptic. Margins: dorsal margin lobed, ventral margin crenate to dentate. Sulcus acusticus: heterosulcoid, ostial, median. Ostium: funnel-like. Cauda: elliptic, straight ending close to the posterior margin. Anterior region: peakeddouble peaked; rostrum long, narrow, pointed; antirostrum long, very narrow, pointed; excisura wide with a very deep acute notch. Posterior region: oblique-lobed.

Siganus sutor (Valenciennes, 1828)

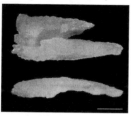 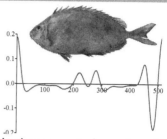

Shape: squared-elliptic. Margins: dorsal margin sinuate, ventral margin entire. Sulcus acusticus: heterosulcoid, ostial, median. Ostium: funnel-like. Cauda: elliptic, markadely curved ending close the ventral margin. Anterior region: double peaked; rostrum long, narrow, pointed; antirostrum long, narrow, pointed; excisura wide with a very deep acute notch. Posterior region: flatened with identations.

Family Acanthuridae

Acanthurus sohal (Forsskål, 1775)

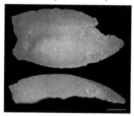 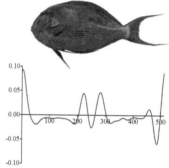

Shape: squared-elliptic. Margins: margins entire to sinuate. Sulcus acusticus: heterosulcoid, ostial, median. Ostium: funnel-like. Cauda: tubular, strongly curved ending close to the ventral margin. Anterior region: peaked; rostrum medium, broad, pointed; antirostrum short, broad, blunt; excisura wide with an acute shallow notch. Posterior region: double-peaked.

Family Sphyraenidae

Sphyraena forsteri Cuvier, 1829

 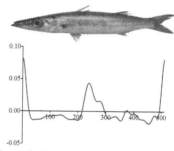

Shape: elliptic. Margins: dorsal margin with a distinct apex ventral margin entire. Sulcus acusticus: heterosulcoid, ostial, median. Ostium: funnel-like. Cauda: tubular, slightly curved ending far from the posterior margin. Anterior region: peaked; rostrum long, broad, pointed; antirostrum poorly defined; excisura wide with a very acute shallow notch. Posterior region: oblique.

Sphyraena jello Cuvier, 1829

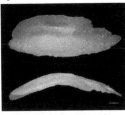
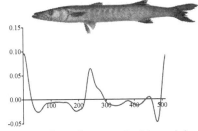

Shape: spindle-shaped. Margins: entire. Sulcus acusticus: heterosulcoid, ostial, median. Ostium: funnel-like. Cauda: tubular, slightly curved ending far from the posterior margin. Anterior region: peaked; rostrum long, broad, pointed; antirostrum short, broad, pointed; excisura wide with an acute shallow notch. Posterior region: oblique with postrostrum lobed.

Sphyraena obtusara Cuvier, 1829

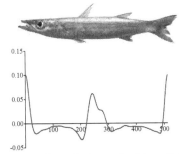

Shape: spindle-shaped. Margins: entire. Sulcus acusticus: heterosulcoid, ostial, median. Ostium: funnel-like. Cauda: tubular, slightly curved ending far from the posterior margin. Anterior region: peaked; rostrum long, broad, pointed; antirostrum short, broad, pointed; excisura wide with an acute shallow notch. Posterior region: oblique.

Sphyraena putnamae Jordan & Seale, 1905

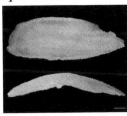
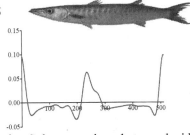

Shape: spindle-shaped. Margins: sinuate to entire. Sulcus acusticus: heterosulcoid, ostial, median. Ostium: funnel-like. Cauda: tubular, slightly curved ending far from the posterior margin. Anterior region: peaked; rostrum long, broad, pointed; antirostrum poorly defined; excisura without notch. Posterior region: oblique with postrostrum pointed.

PERCIFORMES

Family Trichiuridae

Trichiurus lepturus Linnaeus, 1758

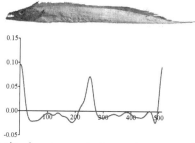

Shape: elliptic-fusiform. Margins: dorsal margin sinuate, ventral margin serrate. Sulcus acusticus: heterosulcoid, ostial, median. Ostium: funnel like. Cauda: tubular, wide straight ending close to the posterior margin. Anterior region: peaked; rostrum long, broad, pointed; antirostrum poorly defined; excisura wide with a square notch. Posterior region: peaked.

Family Scombridae

Auxis thazard thazard (Lacepède, 1800)

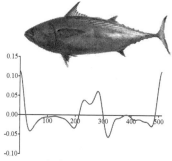

Shape: lanceolated. Margins: ventral margin serrate. Sulcus acusticus: heterosulcoid ostial supramedian. Ostium: funnel like. Cauda: tubular, straight wider posteriorly ending close to the posterior margin. Anterior region: lanceolated; rostrum short, broad, pointed; antirostrum poorly defined; excisura without notch. Posterior region: round-dentate.

Euthynus affinis (Cantor, 1849)

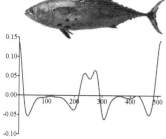

Shape: lanceolated-triangular. Margins: ventral margin serrate. Sulcus acusticus: heterosulcoid, ostial, supramedian. Ostium: funnel like. Cauda: tubular, straight wider posteriorly ending close to the posterior margin. Anterior region: lanceolated; rostrum short, broad, sharp-pointed; antirostrum poorly defined; excisura without notch. Posterior region: oblique-dentate.

Rastrelliger kanagurta (Cuvier, 1816)

 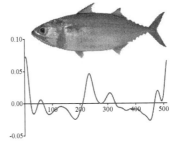

Shape: rectangular with a shallow dorsal concavity. Margins: crenate to dentate. Sulcus acusticus: heterosulcoid, ostial, median with developed ridges. Ostium: funnel like. Cauda: tubular, strongly curved ending close to the posterior margin. Anterior region: peaked; rostrum short, narrow, pointed; antirostrum long, broad, pointed colliculum ostii expanded; excisura wide with a deep acute notch. Posterior region: oblique.

Scomberomorus commerson (Lacepède, 1800)

 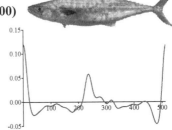

Shape: lanceolated. Margins: serrate. Sulcus acusticus: heterosulcoid, ostial, median with developed ridges. Ostium: funnel like. Cauda: tubular, straight wider posteriorly ending close to the posterior margin. Anterior region: peaked; rostrum long, medium, sharp-pointed; antirostrum medium, narrow, pointed; excisura wide with a deep acute notch. Posterior region: oblique.

Scomberomorus guttatus (Bloch & Schneider, 1801)

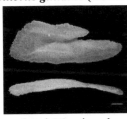 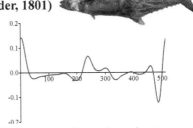

Shape: lanceolated. Margins: dorsal margin sinuate, ventral margin entire to serrate anteriorly. Sulcus acusticus: heterosulcoid, ostial, median with developed ridges. Ostium: funnel like. Cauda: tubular, straight wider posteriorly ending close to the posterior margin. Anterior region: peaked; rostrum long, medium, pointed; antirostrum medium, narrow, pointed; excisura wide with a deep acute notch. Posterior region: oblique-crenate.

Thunnus tonggol (Bleeker, 1851)

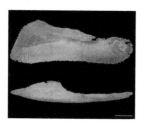
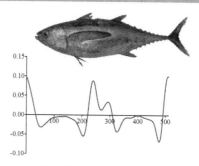

Shape: lanceolated. Margins: serrate. Sulcus acusticus: heterosulcoid, ostial, median with developed ridges. Ostium: funnel like. Cauda: tubular, straight wider posteriorly ending close to the posterior margin. Anterior region: peaked; rostrum long, medium, round; antirostrum medium narrow, pointed; excisura wide with a deep acute dentate notch. Posterior region: oblique.

Family Stromateidae

Pampus argenteus (Euphrasen, 1788)

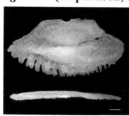
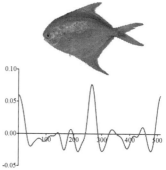

Shape: fusiform. Margins: dorsal margin with medial concavity, ventral margin with deep identations. Sulcus acusticus: heterosulcoid, pseudo-ostial, median. Ostium: oval, perforated colliculum. Cauda: elliptic, straight wider posteriorly ending close to the posterior margin. Anterior region: round; rostrum short, broad, round; antirostrum poorly defined; excisura without notch. Posterior region: angled-oblique.

Family Psettodidae

Psettodes erumei (Bloch & Schneider, 1801)

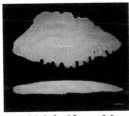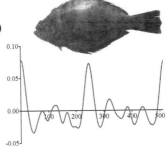

Shape: trapezoidal-fusiform. Margins: dorasal margin lobed, ventral margin with identations. Sulcus acusticus: heterosulcoid, ostial, median. Ostium: funnel like. Cauda: tubular, strongly curved ending far from the posterior margin. Anterior region: peaked; rostrum long, broad, pointed; antirostrum poorly defined; excisura without notch. Posterior region: angled.

Family Paralichthyidae

Pseudorhombus elevatus Ogilby, 1912

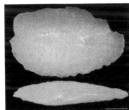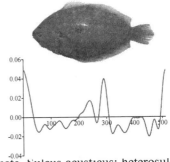

Shape: elliptic-oval. Margins: sinuate to crenate. Sulcus acusticus: heterosulcoid, ostial, median. Ostium: funnel like. Cauda: tubular, straight ending far from the posterior margin. Anterior region: round; rostrum short, broad, round; antirostrum short, broad, pointed; excisura wide with a shallow notch. Posterior region: flatened, with an apex in the middle.

Family Cynoglossidae

Cynoglossus bilineatus (Lacepède, 1802)

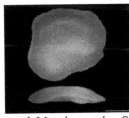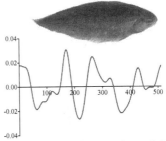

Shape: squared. Margins: entire. Sulcus acusticus: heterulcoid, pseuedo-ostial, median. Ostium: round-oval. Cauda: round-oval, wider than ostium. Anterior region: round; rostrum very short, broad, round; antirostrum absent; excisura absent. Posterior region: flattened.

Fish and otolith measurements

Species	TL	SL	TW	OA	OP	OL	OH
Acanthopagrus berda	340	270	784	40.5	32.8	9.9	6.2
Acanthopagrus bifasciatus	238	184	280	19.9	21.9	7.2	4.0
Acanthopagrus latus	271	205	348	33.9	32.8	9.2	5.6
Acanthurus sohal	390	243	665	13.9	17.5	6.2	3.3
Alectis ciliaris	503	203	378	5.7	17.9	4.9	1.9
Alectis indicus	400	300	629	11.8	18.3	6.3	2.9
Alepes vari	480	344	750	16.4	26.9	8.2	2.7
Anodontostoma chacunda	165	126	71	4.4	21.1	3.5	1.8
Argyrops spinifer	368	286	789	57.0	40.3	12.1	7.2
Atropus atropos	247	185	212	5.4	12.1	4.1	1.9
Auxis thazard thazard	630	550	3200	5.6	16.4	5.0	1.7
Carangoides armatus	380	295	777	12.0	17.6	6.4	2.8
Carangoides bajad	347	279	599	8.3	20.8	5.4	2.2
Carangoides chrysophrys	790	650	5500	25.0	34.3	9.8	4.7
Carangoides malabaricus	219	167	128	7.6	14.2	4.8	2.3
Caranx sexfasciatus	408	315	792	19.4	31.1	8.7	3.2
Cephalopholis formosa	235	192	190	19.4	23.9	7.8	3.6
Cephalopholis hemistiktos	291	234	329	27.7	25.3	8.8	4.3
Chanos chanos	211	152	58	9.1	18.2	6.7	2.3
Cheilinus lunulatus	235	195	275	7.9	15.7	4.4	2.7
Chirocentrus dorab	467	380	270	8.3	17.5	4.6	2.8
Chirocentrus nudus	458	369	295	6.5	12.8	3.9	2.4
Cynoglossus bilineatus	256	-	101	7.6	12.2	3.6	3.0
Decapterus tabl	192	154	67	9.2	18.4	5.2	2.7
Diagramma pictum	660	545	3470	108.3	49.4	16.4	9.2
Drepane longimana	331	261	930	51.7	36.2	10.8	6.7
Drepane punctata	228	185	411	46.5	31.9	10.6	6.3
Dussumieria acuta	132	103	23	4.2	10.1	3.3	1.8
Eleutheronema tetradactylum	354	260	386	19.9	23.5	8.1	3.6
Ephippus orbis	162	130	155	14.1	20.9	5.5	3.7
Epinephelus bleekeri	199	160	92	15.8	20.1	7.2	3.3
Epinephelus caeruleopunctatus	474	382	1750	36.3	31.6	10.5	4.8
Epinephelus chlorostigma	339	169	451	33.2	29.0	10.2	4.9
Epinephelus coioides	342	285	506	37.0	30.1	11.0	4.8
Epinephelus diacanthus	245	195	185	39.3	29.6	10.8	5.3
Epinephelus epistictus	325	262	433	53.5	40.6	13.4	6.1
Epinephelus latifasciatus	325	260	442	27.1	30.1	9.6	4.3
Euthynus affinis	445	380	1300	3.0	11.3	4.1	1.2
Gerres filamentosus	218	158	131	24.2	22.8	7.8	4.8
Gerres longirostris	277	203	269	23.1	27.0	7.1	4.5
Gnathanodon speciosus	242	176	186	5.4	13.3	4.3	1.9
Heniochus acuminatus	260	168	332	12.7	17.3	5.3	3.4
Herklostichthys lossei	81	63	5	1.7	6.2	2.1	1.2
Ilisha megaloptera	278	210	150	10.5	21.4	5.6	2.9
Leiognathus fasciatus	165	131	66	8.1	14.5	4.6	2.7

Fish and otolith measurements (Cont.)

Species	TL	SL	TW	OA	OP	OL	OH
Lethrinus microdon	360	275	512	26.1	24.3	7.9	4.8
Lethrinus nebulosus	351	261	570	52.0	40.9	10.8	7.0
Lobotes surinamensis	339	276	800	29.6	31.0	8.2	5.2
Lutjanus argentimaculatus	423	345	1280	70.0	37.5	12.7	7.8
Lutjanus bengalensis	210	156	91	27.3	23.5	7.9	5.0
Lutjanus ehrenbergii	146	115	38	18.2	18.9	6.5	3.8
Lutjanus erythropterus	316	242	438	61.6	34.1	11.7	7.7
Lutjanus fulviflamma	176	136	78	22.1	20.5	7.3	4.3
Lutjanus johnii	167	131	71	30.3	24.1	8.5	5.1
Lutjanus lemniscatus	298	228	356	44.5	29.2	10.2	6.1
Lutjanus lutjanus	153	123	47	19.9	20.0	6.9	4.1
Lutjanus malabaricus	235	182	210	55.7	31.7	10.9	7.5
Lutjanus quinquelineatus	234	180	188	39.0	33.4	9.3	6.1
Lutjanus rivulatus	405	320	1465	105.2	49.4	15.2	9.5
Lutjanus russellii	150	117	47	16.0	19.5	6.6	3.7
Megalaspis cordyla	486	401	1260	13.4	25.4	8.7	2.3
Mene maculata	178	140	106	5.2	14.1	3.9	2.3
Monodactylus argenteus	200	148	172	21.7	22.1	7.1	4.4
Nematolosa nasus	205	157	106	4.8	13.7	4.0	1.7
Nemipterus japonicus	284	194	204	39.2	29.0	9.8	6.0
Nemipterus peronii	263	190	181	21.1	23.4	8.0	4.1
Otolithes ruber	295	25	300	68.2	33.5	11.9	8.2
Pampus argenteus	340	240	717	39.5	47.9	10.8	5.3
Parastromateus niger	501	375	2324	17.8	24.7	8.6	3.8
Parupeneus rubescens	230	179	139	4.9	15.1	3.1	2.4
Pennahia macrophthalmus	215	170	165	49.3	29.3	9.0	7.2
Pinjalo pinjalo	210	154	130	22.7	21.8	7.4	4.5
Platax orbicularis	442	345	3120	20.9	23.8	7.1	3.9
Platycephalus indicus	352	307	260	21.4	28.0	10.0	3.3
Plectorhinchus flavomaculatus	244	192	192	26.8	30.4	8.1	4.6
Plectorhinchus gaterinus	300	240	443	27.5	24.8	8.5	4.5
Plectorhinchus pictus	421	344	1252	39.7	33.4	10.2	5.6
Plectorhinchus schotaf	301	235	450	18.9	21.1	7.2	3.8
Pomacanthus maculosus	380	360	933	41.5	28.4	10.1	5.8
Pomadasys argenteus	475	375	1452	134.1	48.8	16.1	11.8
Pomadasys kaakan	374	316	696	163.1	54.3	18.4	11.9
Priacanthus blochii	452	354	1325	6.8	12.2	3.8	2.9
Priacanthus tayenus	199	151	117	5.3	13.7	3.1	2.6
Pristipomoides sieboldii	235	175	139	29.6	30.0	8.9	5.7
Psettodes erumei	399	321	825	41.4	30.8	11.1	5.9
Pseudorhombus elevatus	268	210	211	12.6	15.3	5.3	3.2
Rachycentron canadum	605	470	1475	8.7	20.1	6.2	1.9
Rastrelliger kanagurta	268	215	210	8.5	19.9	5.2	2.6
Rhabdosargus haffara	177	132	95	14.6	20.5	5.6	3.7
Sardinella gibbosa	135	109	18	2.2	7.4	2.6	1.2

Fish and otolith measurements (Cont.)

Species	TL	SL	TW	OA	OP	OL	OH
Saurida tumbil	461	382	680	30.9	32.1	11.8	3.8
Scarus ghobban	325	260	564	12.0	18.7	5.4	3.3
Scarus persicus	322	255	573	54.0	32.3	11.5	6.8
Scolopsis ghanam	191	148	100	8.1	15.8	4.9	2.7
Scolopsis taeniatus	235	174	169	8.6	21.6	4.6	2.8
Scolopsis vosmeri	168	131	92	7.6	14.2	4.7	2.3
Scomberoides commerson	980	780	6540	24.6	31.6	12.2	3.7
Scomberoides tol	333	270	201	5.4	19.4	4.9	1.8
Scomberomorus commerson	298	238	157	7.5	16.1	5.6	2.2
Scomberomorus guttatus	530	420	1105	29.0	45.7	10.7	4.2
Selar crumenophthalmus	218	166	128	8.8	14.3	4.4	2.8
Selaroides leptolepis	136	118	27	4.7	10.3	3.5	1.9
Seriola dumerili	499	380	1308	10.7	20.8	7.0	2.5
Seriola rivoliana	401	306	770	6.8	21.1	5.0	2.2
Siganus javus	272	206	225	5.7	15.4	4.2	2.1
Siganus sutor	338	258	590	5.4	16.6	4.4	1.9
Sillago sihama	225	185	74	21.6	19.6	7.2	4.1
Sphyraena forsteri	267	213	106	17.1	20.9	7.2	3.4
Sphyraena jello	760	-	1628	46.8	36.9	13.2	4.9
Sphyraena obtusata	220	182	63	14.9	19.5	7.5	3.0
Sphyraena putnamae	500	412	555	30.7	29.5	10.6	3.9
Terapon jarbua	275	216	279	24.6	23.7	8.0	4.8
Terapon theraps	167	136	72	26.2	22.5	7.3	5.1
Thunnus tonggol	475	405	1303	11.9	25.7	7.1	2.8
Trachinotus mookalee	363	250	700	9.2	19.4	6.0	2.3
Trichiurus lepturus	990	390	690	16.3	21.0	7.9	3.0
Tylosurus crocodilus	1120	1050	2300	34.2	28.2	9.3	4.8
Upeneus sulphureus	167	159	59	8.7	13.7	4.3	3.0
Upeneus tragula	221	179	156	7.2	13.5	4.5	2.4

References

AFORO website. http://www.cmima.csic.es/aforo.
Assis, C.A. 2000. Estudo morfológico dos otólitos sagitta, ssteriscus e lapillus de Teleósteos (Actinopterygii, Teleostei) de Portugal continental. Sua aplicação em estudos de filogenia, fistemática e ecologia. PhD thesis, University of Lisbon.
Assis, C.A. 2004. Guia para a identificação de algumas famílias de peixes ósseos de Portugal continental, através da morfologia dos seus otólitos sagitta. Câmara Municipal de Cascais, Cascais.
Beech, M. J. 2004. The fish fauna of Abu Dhabi Emirate. In: Loughland RA, Al Muhairi FS, Fadel SS, Al Mehdi AM, Hellyer P (eds) Marine atlas of Abu Dhabi. Emirates Heritage Club, Abu Dhabi.
Blacker, R.W. 1969. Chemical composition of the zones in cod (Gadus morhua L.) otoliths. J. Cons. Int. Explor. Mer 33: 107-108.
Bonhomme, V., Picq, S., Claude, J. 2012. Momocs: Shape Analysis of Outlines. R package version 0.2-01. http://CRAN.R-project.org/package=Momocs.
Brewer, P. G, Dyrssen, D. 1985. Chemical oceanography of the Persian Gulf. Prog. Oceano. 14: 41–55.
Campana, S.E. 2004. Photographic atlas of fish otoliths of the Northwest Atlantic ocean. Can. Spec. Publ. Fish. Aquat. Sci. 133: 1-284.
Carlström, D. 1963. A crystallographic study of vertebrate otoliths. Biol. Bull., 125: 441-63.
Cermeño, P., Morales-Nin, B., Uriarte, A. 2006. Juvenile European anchovy otolith microstructure. Sci. Mar. 70: 553-57.
Chao, S. Y, Kao, T. W, Al-Hajri, K. R. 1992. A numerical investigation of circulation in the Arabian Gulf. J. Geophys. Res. 7: 11219–11236.
Carpenter, K. E., Krupp, F., Jones, D. A., Zajonz, U. 1997. Living marine resources of Kuwait, Eastern Saudi Arabia, Bahrain, Qatar and UAE. FAO Species Identication Field guide for Fishery Purposes.
Claude, J. 2008. Morphometrics with R. Use R! New York: Springer.
Degens, E. T., Deuser, W. G., Haedrich, R. L. 1969. Molecular structure and composition of fish otoliths. Mar. Biol. 2: 105-13.
Eschmeyer, W. N, Ferraris, C. J. Jr., Hoang, M. D., Long, D. J. 1998. Part I. Species of fishes. In: W.N. Eschemeyer (ed.), Catalog of fishes. California Academy of Sciences Special publications, San Francisco.
Fitch, J. E., Brownell, R.L. 1968. Fish otoliths in cetacean stomachs and their importance in interpreting feeding habits. J. Fish. Res. Bd Can. 25: 2561-2574.
Furlani, D, Gales, R., Pemberton. D. 2007. Otoliths of common Australian temperate fish. A photographic Guide. CSIRO Publishing, Collingwood.
García-Godos, I. 2001. Patrones morfológicos del otolito sagitta de algunos peces óseos del Mar Peruano. Bol. Inst. Mar Perú 20: 1-83.
Härkönen, T. 1986. Guide to the otoliths of the bony fishes of the Northeast Atlantic. Danbiu ApS., Hellerup.
Hecht, T. 1987. A guide to the otoliths of Southern Ocean fishes. S. Afr. J. Antarct. Res. 17: 1-87.
Kämpf, J, Sadrinasab, M. 2006. The circulation of the Persian Gulf: A numerical study. Ocean Science. 2: 27–41.
Kardovani, P. 1995. Iranian marine ecosystem (The Persian Gulf and the Caspian Sea) Tehran, Iran: Ghomes.

Mallat, S. 1991. Zero-crossings of a wavelet transform. IEEE Transactions on Information Theory 37: 1019-1033.

Morrow, J. E. 1977. Ilustrated keys to otoliths of forage fishes of the Gulf of Alaska, Bering Sea and Beaufort Sea. In: Environmental Assessment of the Alaskan Continental Shelf, pp.757-825. Outer Continental Shelf Environmental Assessment Program, Boulder, Colorado

Morrow, J. E. 1979. Preliminary keys to otoliths of some adult fishes of the Gulf of Alaska, Bering Sea and Beaufort Sea. NOAA Tech. Rep. 1-420.

Nelson, J. S. 2006. Fishes of the world. 4th ed. John Wiley and Sons, Hoboken, New Jersey.

Nolf, D. 1985. Otolithi piscium. in: Schultze, HP, editor. Handbook of paleoichthyology. New York: Gustav Fisher Verlag.

Parisi-Baradad, V., Lombarte, A., García-Ladona, E., Cabestany, J., Piera, J., Chic, O. 2005. Otolith shape contour analysis using affine transformation invariant wavelet transforms and curvature scale space representation. Mar. Fresh. Res. 56: 795-804.

Parisi-Baradad, V, Manjabacas, A, Lombarte, A, Olivella, R, Chic, Ò, Piera, J., García-Ladona, E. 2010. Automatic Taxon Identification of Teleost fishes in an otolith online database. Fish. Res. 105: 13-20.

Platt, C, Popper, A. N. 1981. Fine structure and function of the ear. In W. N. Tavolga, A. N. Popper, and R. R. Fay (eds.). Hearing and sound communication in fishes, pp. 3-38. Springer-Verlag. New York.

Popper, A. N, Fay, R. R. 1993. Sound detection and processing by fish: critical review and major research questions. Brain Behav. Evol. 41: 14-38.

Price, A. R. G, Sheppard, C. R. C, Roberts, C. M. 1993. The Gulf: its biological setting. Mar. Poll. Bull. 27: 9–15.

Reynolds, M. R. 1993. Physical oceanography of the Gulf, Strait of Hormuz, and Gulf of Oman—results from the Mt. Mitchell expedition. Mar. Poll. Bull. 27:35–39.

Rivaton, J., Bourret, P. 1999. Les otolithes des poissons de l'Indo-Pacifique. Doc. Sci. Tech. II 2: 1-378.

Sadighzadeh, Z., Otero-Ferrer, J. L., Lombarte, A., Fatemi, M. R., Tuset, V. M. in press. Otolith shape as an indicator of functional and ecological role of fishes in ecosystems. Mar. Fresh. Res.

Schmidt, W. 1968. Vergleichend morphologische Studie über die Otolithen mariner Knochenfische. Archiv für Fischereiwissenschaft, 19: 1-96

Schmidt, W. 1969. The otoliths as a means for differentiation between species of fish of very similar appearance. Proc. Symp. Oceanog. Fish. Res. Trop. Atl., UNESCO, FAO, OAU.

Sheppard, C., Price, A., Roberts, C. 1992. Marine Ecology of the Arabian Region. Patterns and processes in extreme tropical environments. Academic Press.

Sheppard, C. R. C. 1993. Physical environment of the gulf relevant to marine pollution: an overview. Mar. Poll. Bull. 27: 3-8.

Sheppard C., Loughland, R. 2002. Coral mortality and recovery in response to increasing temperature in the southern Arabian Gulf. Aquat. Ecosyst. Health Manage. 5: 395- 402.

Smale, M. J, Watson G., Hecht, T. 1995. Otolith atlas of Southern African marine fishes. Ichthyol. Monogr. JLB. Smith Inst. Ichthyol. 1: 1-356.

Tuset V. M., Lombarte, A., Assis, C. 2008. Otolith atlas for the western Mediterranean, north and central eastern Atlantic. Sci. Mar. 72S1: 7-198.

Valinassab, T., Daryanabard, R., Dehghani, R., Pierce, G. J. 2006. Abundance of demersal fish resources in the Persian Gulf and Oman Sea. J. Mar. Biol. Assoc. U.K 86: 1455-62.

Valinassab,T., Adjeer, M., Momeni, M. 2010. Biomass estimation of demersal fishes in the

Persian Gulf and Oman Sea by swept area method. Final report (in Persian). Iranian Fisheries Research Organization Press.

Van Lavieren, H., Burt, J., Feary, D. A., Cavalcante, G., Marquis, E., Benedetti, L., Trick, C., Kjerfve, B., Sale, P. F. 2011. Managing the growing impacts of development on fragile coastal and marine ecosystems: Lessons from the Gulf. A policy report, UNU-INWEH, Hamilton, ON, Canada.

Volpedo, A. V., Echeverría, D. D. 2000. Catálogo y claves de otolitos para la identificación de peces del Mar Argentino. 1. Peces de Importância Econômica. Editorial Dunken, Buenos Aires.

Williams, R., McEldowney, A. 1990. A guide to the fish otoliths from waters off the Australian Antarctic Territory, Heard Macquarie Islands. Anare Res. Notes 75: 1-173.

Acknowledgements

We wish to thank all the people having contributed to this work, including our families.

We would like to express our gratitude to Dr. Motallebi the head of Iranian Fisheries Research Organization for his encouragements and providing supports for residence in south of Iran and for making available the technical demands, endless thanks to Institut de Ciències del Mar (CSIC) for scientific support and to Organogenesis Laboratory of Tabriz University for providing steriomicroscope imaging equipments.

Very special thanks to Professors from Islamic Azad University, Marine Biology Department, Science and Research Branch of Tehran Gholamhossein Vosugi, Prof. Dr. Seiied Mohammad Reza Fatemi and Prof. Dr. Tooraj Valinassab for their valuble advises. Thanks to Dr. Ali Salarpouri from Persian Gulf & Oman Sea Ecological Research Institute for providing images of sampling gears (Cages) and maps.

Dr. Victor Tuset is worker of CSIC within the modality «JAE-Postdoc» of Programme «Junta para la Ampliación de Estudios» co-funded by the European Social Foundation (FSE).

Authors

i want morebooks!

Buy your books fast and straightforward online - at one of world's fastest growing online book stores! Environmentally sound due to Print-on-Demand technologies.

Buy your books online at
www.get-morebooks.com

Kaufen Sie Ihre Bücher schnell und unkompliziert online – auf einer der am schnellsten wachsenden Buchhandelsplattformen weltweit! Dank Print-On-Demand umwelt- und ressourcenschonend produziert.

Bücher schneller online kaufen
www.morebooks.de

VDM Verlagsservicegesellschaft mbH
Heinrich-Böcking-Str. 6-8   Telefon: +49 681 3720 174   info@vdm-vsg.de
D - 66121 Saarbrücken   Telefax: +49 681 3720 1749   www.vdm-vsg.de

Made in the USA
Middletown, DE
17 October 2022